D0948555

Bound Heckman · 8/11/78

47-42 70E 3.75

STATISTICS:
A Guide to Business and Economics

HOLDEN-DAY SERIES IN PROBABILITY AND STATISTICS

Erich L. Lehmann, Editor

Bickel, Peter J., and K.A. Doksum: *Mathematical Statistics**
Carlson, Roger: *Introduction to Statistics*
Freedman, David: *Approximating Countable Markov Chains*
Freedman, David: *Brownian Motion and Diffusion*
Freedman, David: *Markov Chains*
Hajek, Jaroslav: *Nonparametric Statistics*
Hodges, J. L., Jr., and E. L. Lehmann: *Basic Concepts of Probability and Statistics, 2d ed.*
Hodges, J. L., Jr., and E. L. Lehmann: *Elements of Finite Probability, 2d ed.*
Lehmann, E. L.: *Statistical Methods Based on Ranks*
Nemenyi, Peter, and Sylvia K. Dixon: *Statistics From Scratch*
Neveu, Jacques: *Mathematical Foundations of the Calculus of Probability*
Parzen, Emanuel: *Stochastic Processes*
Renyi, Alfred: *Foundations of Probability*
Tanur, Judith M., Frederick Mosteller, William Kruskal, et al: *Statistics: A Guide to the Unknown*
Tanur, Judith M., Frederick Mosteller, William Kruskal, et al: *Statistics: A Guide to Business and Economics*

**To be published*

STATISTICS: A Guide to Business & Economics

edited by the editors of
STATISTICS: A GUIDE TO THE UNKNOWN

JUDITH M. TANUR
State University of New York, Stony Brook

and **FREDERICK MOSTELLER, Chairman**
Harvard University

WILLIAM H. KRUSKAL
University of Chicago

RICHARD F. LINK
Artronic Information Systems, Inc.
and Princeton University

RICHARD S. PIETERS
Phillips Academy, Andover, Mass.

GERALD R. RISING
State University of New York, Buffalo

**The Joint Committee on
The Curriculum in Statistics and Probability of
The American Statistical Association and
The National Council of Teachers of Mathematics**

and
by **E. L. LEHMANN**
University of California, Berkeley
Special Editor for Statistics: A Guide to Business & Economics

hd HOLDEN-DAY, INC.

SAN FRANCISCO London Dusseldorf Johannesburg
Panama Singapore Sydney

Copyright © 1976 by Holden-Day, Inc.
500 Sansome Street, San Francisco, California 94111

All rights reserved

No part of this book may be reproduced, stored in a retrieval system, or transmitted, in any form or by any means, electronic, mechanical, photocopying, recording, or otherwise, without permission in writing from the publisher

ISBN 0-8162-8614-0

Library of Congress Catalog Card Number 76-5708

Printed in the United States of America

PREFACE

WARREN WEAVER, a great expositor of science, discussed why science is not more widely appreciated and issued a call in "The Imperfections of Science" (*American Scientist,* 49:113, March 1961):

> What we must do—scientists and non-scientists alike—is close the gap. We must bring science back into life as a human enterprise, an enterprise that has at its core the uncertainty, the flexibility, the subjectivity, the sweet unreasonableness, the dependence upon creativity and faith which permit it, when properly understood, to take its place as a friendly and understanding companion to all the rest of life.

As this book makes clear, scientific thinking, most particularly related to statistics, is not restricted to "pure" science, but has many uses in business and economics as well. And most especially, it is necessary to develop what we might call a statistical attitude toward, and manner of thinking about, these disciplines.

To prepare a volume describing important applications of statistics and probability in many fields of endeavor—this was the project that the ASA-NCTM Committee initiated early in 1969 as an effort to close the gap Weaver and others had pointed out. *Statistics: A Guide to the Unknown* (SAGTU) was the result.

During the book's preparation several of us who were working on it and teaching simultaneously found much of the material very useful—even inspirational—to undergraduate and graduate students. It seemed that the book had an additional possible function as an auxiliary textbook. This impression has been confirmed over the years since publication of *SAGTU* as college after college, university after university, and even secondary after secondary school adopted *SAGTU* as an auxiliary textbook for introductory statistics classes.

Instructors and students have reported success in using *SAGTU* as a means of tieing techniques and methods, taught necessarily at a simple level with simplified examples, to real problems in the real world. In specialized courses, some teachers wanted sets of essays oriented to their subject matter. Students studying business and economics, for example, found themselves only distantly concerned with statistical applications in biologic sciences. The very diversity of applications that had fascinated us became an impediment to the usefulness of *SAGTU* as an auxiliary textbook within the time constraints of a specialized statistics course. It is for this reason that the decision was taken to compile what we have come to refer to as mini-*SAGTU's*: each a selection of articles dealing with a particu-

790568 LIBRARY
ALMA COLLEGE
ALMA, MICHIGAN

lar field of application—*Statistics: A Guide to Business and Economics* (SAGBE) is the first such volume to appear.

The essays here, and in SAGTU itself, do not teach technical methods, but rather illustrate past accomplishments and current uses of statistics and probability. In choosing the actual essays to include, the Committee aimed at illustrating a variety of applications, but did not attempt the impossible task of covering all possible uses. Even in the areas included, attempts at complete coverage have been deliberately avoided. We discouraged authors from writing essays that could be entitled "All Uses of Statistics in . . ." Rather, we asked authors to stress one or a very few important problems within their field of application and to explain how statistics and probability help to solve them and why the solutions are useful to the nation, to science, or to the people who originally posed the problem. In the past, for those who were unable to cope with very technical material, such essays have been hard to find.

When describing work in the mathematical sciences, one must make a major decision as to what level of mathematics to ask of the reader. Although the Joint Committee serves professional organizations whose subject matter is strongly mathematical, we decided to explain statistical ideas and contributions without dwelling on their mathematical aspects. This was a bold stroke, and our authors were surprised that we largely held firm.

There is an old saw that a camel is a horse put together by a committee. Our authors supplied exceedingly well-formed and attractive anatomical parts, but to the extent that this book gaits well, credit is due primarily to a most talented and dedicated Committee. In general, the approach to unanimity in the Committee's critical reviews of and suggestions about essays was phenomenal. And, though they may have occasionally been divided about the strong and weak points of a particular essay, they were constantly united in their purpose of producing a useful book, and in their ability to find something more than 24 hours a day to work on it. This dedication undoubtedly created difficulties for our authors. Nevertheless, our authors persevered and deserve enormous thanks from us, and from the Committee, and from the statistical profession at large.

Our thanks go also to the Sloan Foundation whose grant made it possible to put this book together.

There are others to thank as well: to the office of the American Statistical Association (and, in particular, to Edgar Bisgyer, and John Lehman, and later Fred Leone) for invaluable help in all the administrative work necessary to get out a book such as this; and similar thanks to the administration of the National Council of Teachers of Mathematics; to Edward Millman for careful and imaginative editorial assistance; and to other people at Holden-Day, especially Frederick H. Murphy, and Erich Lehmann, our Series Editor; to Mrs. Holly Grano for acting as a long-distance and long-haul secretary; and to the many friends and colleagues both of the Editor and of the Committee members who so often acted as unsung, but indispensable, advisors.

In this new effort to compile *SAGBE*, additional thanks go to the original Committee members and to the guiding spirit of Erich Lehmann. The supplementary material for study, prepared especially for *SAGBE* is the valuable contribution of David Lane and Rick Persons.

Frederick Mosteller
Judith Tanur

Atlanta, Georgia
October, 1975

FOREWORD

The Right Honorable Harold Wilson, Prime Minister of Great Britain, in opening the 37th Session of the International Statistical Institute in London, September 4, 1969, said:

> The list of papers for the Session reflects the ever widening range of application of statistical methods. When I joined the Royal Statistical Society the papers read were still mainly on economic and social statistics. Nowadays the papers read before a society like those for your session of the Institute cover many more topics relating to many disciplines. It means, I am afraid, that as statisticians today you help so many people in so many diverse subject fields that none of your clients can see the overall contribution which you as statisticians make together as a whole. As a result your value is not perhaps sufficiently recognized by any one group of the people with whom you deal nor are your great services fully realized by the general public.

Himself a statistician, the Prime Minister understood well these contributions of statistics. He might be interpreted as calling for statisticians to explain what they do.

Warren Weaver, a great expositor of science, discussed why science is not more widely appreciated and issued a similar call in "The Imperfections of Science" (*American Scientist,* 49:113, March 1961):

> What we must do—scientists and non-scientists alike—is close the gap. We must bring science back into life as a human enterprise, an enterprise that has at its core the uncertainty, the flexibility, the subjectivity, the sweet unreasonableness, the dependence upon creativity and faith which permit it, when properly understood, to take its place as a friendly and understanding companion to all the rest of life.

Dr. Weaver has had a lifelong interest in probability and statistics. He was given the first Archives of Science Award for his contribution to public appreciation of science.

In the U.S., too, government representatives want explanations. For example, Craig Hosmer, a House Member of the House and Senate Committee on Atomic Energy, in discussing the funding of science at a technological conference on March 5, 1968, said: "The scientific community should take greater pains to make clear that its efforts contribute directly and indirectly to the public good."

Thus, these and other important men ask scientists to tell the public about their subject and to explain what contribution science makes to society. Their request is easier made than satisfied. This collection of essays on applications of statistics represents one kind of step toward meeting it.

To find the origins of this work, we might turn back to the great change and advance in mathematics education initiated in 1954 when the Commission on Mathematics of the College Entrance Examination Board brought together, for a sustained study of the curriculum, teachers and administrators of mathematics from several sources: secondary schools, teachers' colleges, and colleges and universities. Prior to that gathering, the several groups of teachers had seldom worked together on the problems of the curriculum. That meeting of minds has developed and continued in many directions; one of its long-run consequences was the establishment of the Joint Committee of the American Statistical Association (ASA) and the National Council of Teachers of Mathematics (NCTM) on the Curriculum in Statistics and Probability. By late 1967, such cooperation between school and college teachers was widespread, and it was easy for Donovan Johnson, then President of NCTM, and me, then President of ASA, to set up the Joint Committee to review matters in the teaching of statistics and probability.

Early in its work the Joint Committee decided that it wanted to encourage the teaching of statistics in schools, for statistics is a part of the mathematical sciences that deals with many practical, as well as esoteric, subjects and is especially organized to treat the uncertainties and complexities of life and society. To explain why more statistics needs to be taught, we need to make clearer to the public what sorts of contributions statisticians make to society. In the field of statistics, we are, indeed, responding to the sort of requests quoted above.

When describing work in the mathematical sciences, one must make a major decision as to what level of mathematics to ask of the reader. Although the Joint Committee serves professional organizations whose subject matter is strongly mathematical, we decided to explain statistical ideas and contributions without dwelling on their mathematical aspects. This was a bold stroke, and our authors were surprised that we largely held firm.

The Joint Committee has been extremely fortunate to find so many distinguished scholars willing to participate in this educational project. The authors' reward is almost entirely in their contribution to the appreciation of statistics. We have been fortunate, too, to have Judith Tanur as editor of the collection and hard working committee members as her staff.

To teachers, I can report that thus far I have used material from about one-third of the essays in classes and in speeches. Adult students seemed to enjoy discussing the data and reading further. We do not, however, regard the book as a textbook. We have had other favorable reports from adults who were not students and who read the articles voluntarily. Perhaps the most heartening report on readability came from one of our authors, whose secretary told him, after finishing the typing of a revision, that she enjoyed it enormously. When asked what she especially liked, she said that she had finally found out what the work of the office was all about.

In a parallel writing effort, the Joint Committee has also produced a series of pamphlets for classroom teaching entitled *Statistics by Example*. Intended for students whose mathematical preparation is modest, these volumes teach statistics by means of real-life examples. That effort differs from this one in that the student learns specific techniques, tools, and concepts by starting from concrete examples. (The publisher is Addison-Wesley, Sand Hill Road, Menlo Park, California 94025.)

Some readers may wish to know how to become statisticians, and others may have the obligation to advise students about career opportunities. The brochure *Careers in Statistics* (obtainable from the American Statistical Association, Suite 640, 806 Fifteenth Street, Washington, D.C. 20005) provides information about the nature of the work and the training required for various statistical specialties.

The Joint Committee appreciated being able to report on its work at ten meetings of the National Council or its affiliates. We also reported to the American Statistical Association at Chicago, Illinois, and Detroit, Michigan; to a conference called by the National Science Foundation at the University of Minnesota; to the International Statistical Institute workshop on teaching statistics at Oisterwijk, Netherlands; and to the international conference on teaching of probability and statistics of the Comprehensive School Mathematics Program at Carbondale, Illinois. I discussed some of the material in one of my Allen T. Craig lectures at the University of Iowa.

In addition, by its existence at Harvard University, National Science Foundation grant GS-2044X2 has considerably facilitated this project without directly supporting it. Much of the work was done during periods while Frederick Mosteller held a Guggenheim Fellowship and while William Kruskal was a National Science Foundation Senior Postdoctoral Fellow at the Center for Advanced Study in the Behavioral Sciences. We have also benefited from a number of courtesies extended by the Russell Sage Foundation and by the Social Science Research Council. Before resigning to take up the tasks of the presidency of the National Council of Teachers of Mathematics, Julius Hlavaty was a member of the Joint Committee and participated in the decision to create this collection. The national offices of ASA and NCTM have been most helpful, as have representatives of our publisher, Holden-Day, Inc.

Finally, we have no monopoly on the task of explaining statistics to the public. We urge others to provide their views on the purposes, the methods, and the results of statistical science.

Frederick Mosteller, Chairman
Joint Committee of ASA–NCTM

Cambridge, Massachusetts
February 14, 1972

CONTENTS

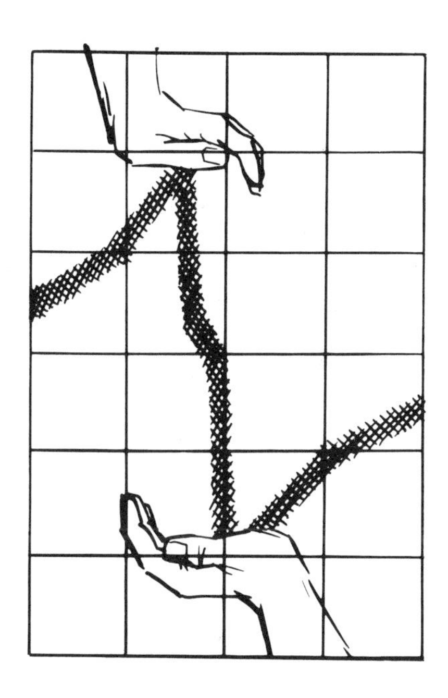

STATISTICS FOR PUBLIC FINANCIAL POLICY

Leonall C. Andersen *Federal Reserve Bank, St. Louis*

ARTHUR BURNS,[1] Chairman of the Federal Reserve Board, sat back in his paneled office on Constitution Avenue, puffed his pipe pensively, and reflected on the forthcoming meeting of the Fed's Open Market Committee. What should they decide about purchase or sale of government securities in the open market, and about the many other instruments of policy available to the Committee?

Whatever the decisions might be, their consequences would affect wages paid to the steel worker in Gary, Ind., and the movie extra in Los Angeles; they would also affect the profits and the expansion plans of the largest steel company and the smallest corner grocer; and they would affect the prices paid for apples, automobiles, and xylophones by everyone.

[1] The names, for concreteness, are those of February 1971, when this was written. They may change, but the issues will continue.

At about the same time, Paul McCracken, Chairman of the Council of Economic Advisors, sat back in his office in the Executive Office Building, a less elegant, but more historic (and hard by the White House) office than that at the Federal Reserve. What should the Council recommend to the President about Federal spending and other fiscal actions? The Council cannot, of course, move in the independent and decisive ways open to the Federal Reserve, but its recommendations and arguments can be highly influential upon the views of the President, the Congress, and all Federal agencies dealing with economic matters.

Whatever the Council's recommendations might be, their implementation too would have effects on wages, profits, prices, and other economic quantities. My pocketbook and yours are linked inextricably to such decisions at the levels of great power and great wealth.[1]

The links, however, are complicated, they change over time, they have many interconnections, and some of their effects are very difficult to measure accurately and quickly. Economics is the science that tries to understand these complex matters.

FISCAL AND MONETARY ACTIONS

To discuss governmental economic policy at all, we must have some idea of goals and some idea of possible actions to try to achieve those goals. Three sorts of goals are widely accepted for the national economy:

(1) High employment
(2) Relatively stable prices
(3) Rising output of goods and services

It might be objected that these three goals contain mutually contradictory elements, and there have long been doubts (sharply increased recently) about working toward a rising output of goods and services, especially as it is usually measured by the Gross National Product (GNP).

Now what about actions? In this essay we shall consider two kinds of actions:

(1) *Monetary actions.* These are actions that change the amount of money and credit available to the economy. For example, a purchase of securities in the nation's money market (Wall Street) by the Federal Reserve will tend to increase the total available money and credit. (You should keep in mind that money includes the total of all checking accounts in banks.) Monetary actions are primarily under the control of the Federal Reserve System, and of the Treasury.

[1] A high official of the Office of Management and Budget recently quipped that his office is the only one in existence where 0.1 means one hundred million dollars.

(2) *Fiscal actions.* These are broad-scale acts of Government spending and taxation. They relate to the much-argued issue of government deficit financing.

In recent years a debate has been hotly raging over monetary versus fiscal actions as the more effective means of achieving important economic goals. The fundamental Keynesian viewpoint that has now become part of the mainstream (President Nixon has been quoted as saying "I am a Keynesian") concentrates almost exclusively on the direct influence of fiscal actions, primarily of government spending. Some economists, of whom Milton Friedman of the University of Chicago is perhaps the best known, argue for monetary actions as more effective and more predictable.

We shall here completely omit some aspects of this debate, in particular the difficult underlying economic theory and such institutional arguments as the ease of taking monetary actions relative to fiscal—the former can be effected by a single agency, while the latter require the whole political process of interaction between the Executive and Legislative segments of the government. (Fiscal proponents might rebut by emphasizing the democratic desirability of having major decisions go through the political process, even at the inherent cost of clumsiness and time lost.) Rather, we shall indicate how statistical methods have permitted insight into a direct empirical comparison of the two kinds of government action.

MEASUREMENTS AND PROCEDURE

An essay of this character, brief and for a wide audience, can hardly hope to provide more than a surface glimpse of a highly technical topic. Yet it is important, I feel, that such glimpses be given frequently and in many ways. Economics, like war, is too important to be left entirely to the specialists.

What we shall describe is a small part of a correlation study. (See the end of the essay for a discussion of the correlation notion).

As an indicator of economic well-being we shall use the Gross National Product. The GNP shows, in a word, the total goods and services sold in the market place during some specified period.

As a measure of monetary actions, we shall use the money stock, defined as currency and demand bank deposits held by nonbank individuals and firms.

As a measure of fiscal actions, we shall use government spending.

These three quantities are known each quarter, and the analysis is based, in fact, on quarter-to-quarter *differences*. Thus, suppose that in a certain quarter

(1) GNP increases by $400 million
(2) Money stock increases by $800 million
(3) Government spending decreases by $50 million.

Then in correlating the money stock and GNP, we would use the pair of numbers (800, 400) for that particular quarter-to-quarter change; in correlating spending and GNP, we would use the pair (—50, 400).

RESULTS

Scatter diagrams showing such relationships for the 68 quarters from 1953 to 1969 are shown in Figure 1. The major point to notice is that changes in GNP and those in the money stock are more closely related than changes in GNP and those in spending.

In fact the computed correlation coefficients are:

Between GNP and the money stock: 0.66
Between GNP and spending: 0.44.

The first of these is clearly larger than the second; this says that we can predict changes in GNP (by straight-line prediction) much better by using changes in the money stock than by using changes in government spending.

Another way of looking at this is in terms of the "scatter" or variability of GNP changes around the straight line that best fits the cluster of data in the left hand graph of Figure 1, as against the corresponding variability in the right-hand graph. The relevant quantities here are the squares of the correlation coefficients, 0.44 for changes in GNP and the money stock, and 0.19 for changes in GNP and government expenditures. If we measure

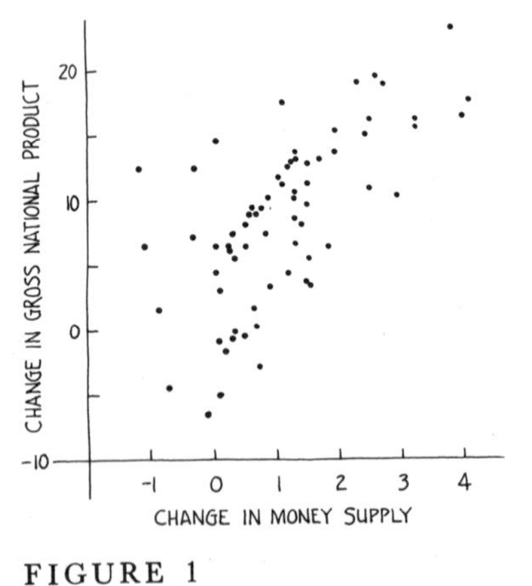

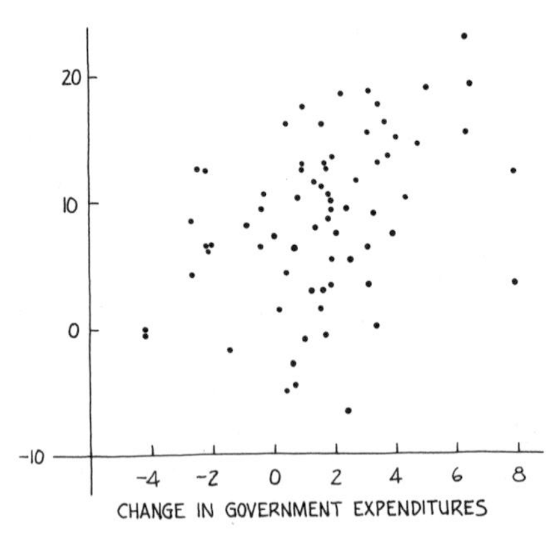

FIGURE 1

Comparison of correlation between changes in gross national product and those in money supply with correlation between changes in GNP and those in government expenditures (figures are in billions of dollars)

variability in terms of proportional decrease in a technical quantity called "variance," then changes in the money stock permit a 44% decrease in variability of changes in GNP, while changes in spending permit only a 19% decrease in that variability.

The full technical analysis was much more detailed and complex than that described above. For example, the full analysis carefully considered time lags: a change in spending now might not affect GNP for two quarters, or the effect of changes in money stock may require four quarters. The refinements, however, did not change the general result: *change in GNP is much more accurately predictable from change in the money stock than from change in spending.*

QUALIFICATIONS

For economic studies, unlike studies in most of the natural sciences, true experiments are generally impossible. We cannot play God, manipulating economic variables this way and that just to see how the economy responds. Economists are as a rule forced to do the best they can with data of the kind discussed earlier—data that arise in the natural course of our changing world.

Under these circumstances it is difficult to establish causation. To say that two quantities are positively correlated is not to say that artificially increasing one would increase the other. For example, imports of gold and the annual number of marriages are positively correlated over years (because both reflect economic health), but suddenly increasing gold imports by government action would scarcely be expected to change the number of marriages.

On the other hand, even when evidence is primarily correlational, people frequently do come to conclusions of causation. A current example is the relationship between smoking and health. When people ascribe causation based on a correlational study there is additional information—usually a reasonable theory of the mechanism behind the effect, an unusually high, consistent, and specific correlation, or some other piece of connective evidence.

A significant part of the debate between those favoring monetary actions and those favoring fiscal ones stems from the causation problem. Other aspects of the debate turn on the accuracy of the data and on possible artifacts of the data.

This debate, and similar ones, are bound to continue. One prediction can safely be made, however: the continuing debate will use statistical tools such as correlation coefficients. There is no choice in empirical quantitative argument. Statistical methods are essential, whether or not they are explicitly described.

A frequently used measure of how well one can predict one variable (e.g. GNP) from another variable (e.g. money stock or spending) by means of a linear equation is the *correlation coefficient**: the closer the correlation coefficient is to +1 or −1 (its largest and smallest possible values) the better the linear prediction will work. If the correlation coefficient is positive, an increase in the first variable is predicted to lead to an increase in the second variable; the two variables are then said to be *positively correlated.* Similarly, if a correlation is negative, an increase in the first variable leads to a predicted decrease in the second variable, and the two variables are then said to be *negatively correlated.*

PROBLEMS

1. Explain the difference between monetary actions and fiscal actions.

2. (a) Roughly how many of the 68 quarters from 1953 to 1969 had *smaller increases* in gross national product than the quarter which had the *greatest decrease* in money supply?

   (b) Roughly how many of these 68 quarters had *greater increases* in gross national product than the quarter which had the *greatest increase* in government expenditures? (Hint: use Figure 1).

3. Would you say that the correlational study in this essay is evidence in favor of fiscal actions over monetary actions? Why or why not?

4. (a) For a number of years in the United States, there was a high correlation between public school teachers salaries and liquor consumption. Do you think that the government could have discouraged liquor consumption by decreasing teachers' salaries?

   (b) Briefly explain the difference between correlation and causation.

5. Suppose the study described in this article had concluded that the correlation between GNP and the money stock was 0.1. How would this have affected the Federal Reserve's policies? What if the correlation was 0.9? −0.9?

6. Suppose you noticed that the number of marriages in a certain year increased by 5% from the previous year. Would you guess that U.S. gold imports increased, decreased, or remained about the same as the last year? Why?

THE PLIGHT OF THE WHALES

D. G. Chapman *University of Washington*

BETWEEN THE end of World War I and 1960, several species of whales in the ocean around the Antarctic continent were the basis of an important industry. These giant mammals, the largest that have ever existed on the earth, were sought for animal oil and, to a lesser extent, meal and meat extract (the latter for human consumption) as well as a myriad of byproducts. In antiquity, whalers went out in small boats and endured great risks to capture such large sources of meat. Men continued to hunt whales in small boats with primitive weapons, as portrayed in *Moby Dick,* until late in the nineteenth century. In the twentieth century, whaling has been highly modernized with explosive harpoons, large ships, and powerful radar-equipped catcher boats, which enable the whaling industry to operate in the stormy and inhospitable oceans next to the Antarctic ice cap.

This area of the world, while unfriendly to man, is very inviting to whales, for during the southern summer the waters bloom with small plants which, in turn, feed myriads of minute animals, known generally as *krill.* Certain species of whales catch these by straining large volumes of water in their huge mouths through sievelike filters called *baleen plates* (hence this group of whales is referred to as baleen whales). These whales have no teeth and do not eat fish or other marine mammals. The largest of the baleen whales, and indeed of all whales, are the blue whales, which may reach a length of 100 feet, though 70 to 80 feet is a more usual size.

BLUE WHALES

Immediately following World War II, Europe and Japan were in desperate need of many things, including animal oil. It was not surprising, therefore, that the number of Antarctic whaling catcher boats increased; furthermore, technologies developed during the War made whaling more efficient. As a result, some conservationists feared that Antarctic whales, particularly the blue whales, would be completely eliminated. Figure 1 shows the annual catch of blue whales in the southern oceans in the decade before the War and in the postwar period to 1960. The basis for concern for the blue whales was easy to document, but the catch of other species was stable or increasing. Some of those associated with the industry suggested reasons other than a decline in population for the decline of the blue whale catch and were re-

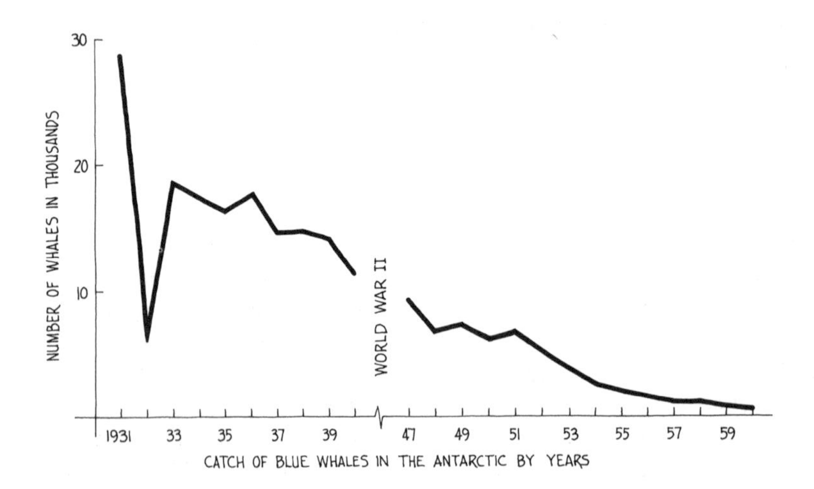

FIGURE 1
Catch of blue whales in the Antarctic, seasons 1930–31 to 1959–60 excluding World War II period

luctant to accept restrictions on catches. Thus the International Whaling Commission, set up in 1946 to manage the resource, found itself the center of controversy. The Commission has representatives from all interested countries; it establishes size regulations and quotas, and also has the authority to ban hunting of species that appear to be endangered.

The Commission set up a study group to bring together all the data and develop statistical methods for attacking such questions as: How many whales are in the stock that feed in the Antarctic? How many young are born each year? How many whales die from natural causes each year? How are these birth and death rates affected by factors over which man has some control?

COUNTING METHODS

Let us consider the first, and perhaps most basic, question: How many whales are there of a particular species? Whales unfortunately don't stay still to be counted. They roam over large areas, spending most of their time under water, though they do surface at regular intervals to breathe. Furthermore, the southern oceans cover a vast part of the world; the whaling area exceeds 10 million square miles, an area larger than all of North America.

Marking Method. There are several standard ways to estimate wild animal populations, all of which involve some statistical techniques. We shall describe three of them. The first involves marking a number of whales; a foot-long metal cylinder is fired into the thick blubber that lies just under the skin. If and when marked whales are later caught, some information is available on their movement, on their rate of capture, and on the proportion of marked members in the whole herd. The usefulness of the latter information is easily seen by simulating such an experiment with a can of marbles. Assume that like the whale population, the number of marbles in the can is unknown. Now pick a few marbles (say, ten) out of the can, mark them, and return them to the can. Next, stir the whole can thoroughly and draw another sample. Count separately the marked and unmarked marbles. If the unmarked ones are four times as numerous in the sample as the marked ones, we reason that the same is true of the whole canful; but because there is a total of ten marked marbles, we infer there are 40 unmarked marbles, or 50 marbles in total.

This simple scheme has been used with many animal populations, though there are many obvious complications in practice, and for whales this is especially true. How do we know, for example, that the metal mark fired into the blubber actually penetrated and did not ricochet off? Did the crew who cut up the captured whale carefully look for the mark—even a foot-long metal cylinder is easy to overlook in cold, stormy working conditions when the volume being cut up is approximately the size of a house. Also, unlike the marbles, whales are born and die over a period of years. All of these complications require refinements and extensions of the simple experiment outlined here. It is necessary to have a series of experiments extending over many years and to use comparative procedures. For example, if a group of whales is marked in year 1 and a group of the same size is marked in year 2, then, after year 2, the ratio of recoveries of whales marked in year 1 to the recoveries of whales marked in year 2 reflects the proportion of marked whales of group 1 that died in the intervening year. These deaths may have been natural or caused by hunters. Moreover, the *ratio* is a valid measure of this mortality because its numerator and denominator are equally affected by the possible errors listed above. Such a comparative study is only one of the several statistical procedures used to analyze whale-marking data.

Catch-per-Day Method. The second estimation method is based on changes in the rate of catching whales. The rate of catching depends mainly on the frequency with which whales are seen, and other things being equal, this depends on their density. Thus the catch per day reflects the density. How can this be translated into absolute numbers? If the change in catch per day is entirely a result of the removal by man, then it is easy to make this translation; if catching 25,000 whales in one season lowers the catch rate for the next season by 10% then at the outset of the first season there must have been 25,000/0.10, or 250,000 whales.

Again, the situation is more complex than this simple example. Whaling ships hunt over a vast area in difficult conditions, so that the catches fluctuate violently. Whaling companies introduce new technology to improve their efficiency. Moreover, we reemphasize that there are other causes of whale mortality, and that there are new births as well; both of these factors must be taken into account in adjusting the population estimate. One way to overcome some of these difficulties is to adjust for changes in efficiency and also to follow the change in catch per day (adjusted) over a period of several seasons. Figure 2 shows the catch per day of blue whales plotted against the cumulative catch by the whaling-factory ships over the seasons 1953–54 to 1962–63 when natural deaths and births were numerically quite small. As more whales were caught, the catch per day went steadily down. This graph suggests that there were only 10,000–12,000 blue whales in 1953 and this number declined to about 1000 in 1963. As pointed out, there are statistical refinements, and the result obtained in this way must be combined with estimates obtained in other ways.

Age Analysis. The catch-per-day method works well with rapidly declining populations, but in other situations, the complications and corrections make it less useful. Still a third method is available, however, that uses the ages of whales. Just as trees have annual rings in their trunks and fish have

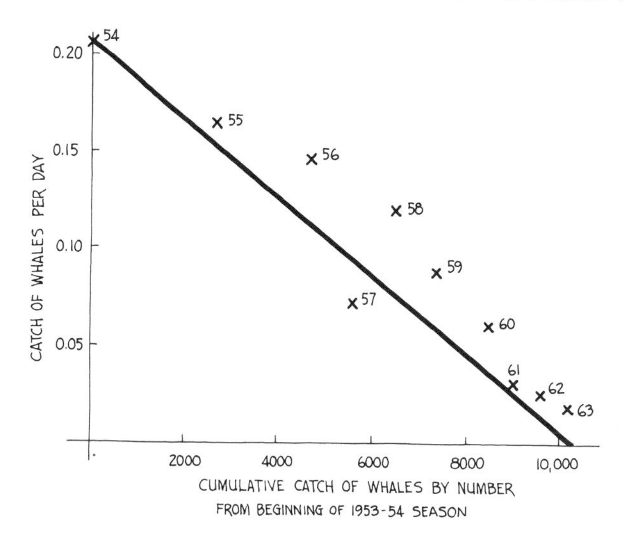

FIGURE 2
Blue whale catch per day (adjusted for efficiency improvements) versus cumulative catch, 1953–54 to 1962–63

annual rings in their scales, whales have annual rings in a waxy secretion in the ear (earplugs). The ages of a sample of the whales killed each year were determined by the rings of their earplugs. In addition, information on the length of every whale killed commercially made it possible to relate age to length and to calculate an estimated age for every captured whale.

It was thus possible to make a statistical estimation of the number of four-year-old whales in any season and the number of five-year-olds in the following season. Because one year's five-year-olds are the survivors of the previous year's four-year-olds, a survival rate or, conversely, a mortality rate can be determined. Because all ages are estimated and because some adjustments have to be made, the estimated mortality rates fluctuate wildly. However, by averaging over several year's classes, over areas and seasons, useful results can be obtained. Furthermore, with additional statistical analysis, it is even possible to assess the magnitudes of possible errors in such estimations. These mortality rates help us to predict the future of the whale population. (See the essay by Keyfitz for an explanation of this method as applied to human populations.)

Results. Thus we have three methods of estimating population sizes and mortality rates: the marking method, the catch-per-day method, and age analysis. The results of different methods were checked against one another, and fortunately the different estimates were in good agreement. Sources of error were carefully checked and ruled out, so that the study group finally concluded that the blue whales numbered at most a few thousand and might total even less than 1000. Thus there was and is a real danger of extinction of this species in the Antarctic (there are also small numbers of blue whales in the northern oceans). Fortunately, the International Whaling Commission banned the taking of blue whales as soon as the study was finished—first, in a large part of the southern oceans and eventually in all waters south of the equator. It is too soon to predict the long-term survival of the species; blue whales are occasionally seen, but these are probably the survivors noted above (whales can live, in the absence of hunting, to more than 40 years of age). We can ask whether the population has been reduced to such low levels that reproduction is reduced below the level necessary for species continuation, but it will be a number of years before this can be answered.

FIN WHALES

The second-largest whale species in the world, also part of the baleen family, is the fin whale. It averages about ten feet less than the blue whale in length. During the fifties this stock annually yielded in excess of one million barrels of oil per year. With the decline of the blue whales, fin whales bore the brunt of the exploitation. The same methods of analysis used for the blue whales were applied to the fin whales; in fact, the analysis was more critically needed because the condition of the fin whale stock was not obvious as was that of the blue whales. Moreover, fin whale catches were still very high: in the 1961–62 season over 27,000 fin whales were killed. The study group recommended that the fin whale catch should be reduced to 7000, or less, if the fin whale stock was not to be further depleted. The proposal for such a drastic reduction came as a shock to the Commission; the study group forecast that the next season's catch, regardless of quotas, would drop to 14,000. When actual figures proved the forecast right, most countries wanted to move toward the drastic reductions required, but some of the whaling nations were able to block action. Another disastrous season caused a revision in the thinking of the commissioners, and in 1965, a substantial schedule of reductions in the quota was agreed upon. Nevertheless, the delay in reaching this agreement and the delay in reducing the quota subsequently, meant that the presently permitted catches have had to be lowered even further. Subsequent analyses using additional data show that, as of 1971, catches in excess of 3000 per year are likely to mean further reduction of the stock. This drastic reduction in the permitted catch has had a severe impact on the industry. Of the five countries that hunted whales with factory ships in the Antarctic in 1960, only Japan and the Soviet Union are still actively engaged in the industry. Further, these have diverted some of their effort to catching smaller whales of little interest in other times and have transferred some factory ships to other oceans or to other types of fishing.

THE FUTURE OF THE WHALE STOCK

The study group was asked to advise the Commission not only on the state of the whale stocks, but also on what the optimum stock size might be. The optimum size is that which will yield the maximum number of whales each year on a continuing basis. It is now recognized that whales should be managed like our fisheries and forestry resources on a sustained-yield basis. So far, no one has found a way of harvesting the plankton production of the southern oceans except via the whales. In 1969, the Soviet Union sent a ship to the Antarctic waters to harvest the krill directly. They succeeded in catching a reasonable quantity, which they converted to a krill paste, reported to be quite tasty. Unfortunately, to meet the costs involved, they were forced to sell it on the Moscow market at about the same price as beef. Faced with a choice between beef and krill paste, the Moscow housewife did exactly what her Western counterpart would do, so the krill harvest was a failure.

The whale stocks should be allowed to increase because the statistical analysis shows that the optimum levels are much higher than the present depleted stock sizes. The steps to permit stocks to increase are difficult, but at least the Commission has now fully accepted the methods of analysis that were first applied to whales in this study. The present quota on catches, not only in southern oceans but also in the North Pacific, is based on the best scientific evidence as reviewed and analyzed by the Scientific Committee of the Whaling Commission, which includes scientists (particularly experts in population statistics) from Canada, Great Britain, Japan, Norway, the Soviet Union, and the U.S., as well as the Food and Agriculture Organization of the United Nations. One loophole remains: each whaling country has the responsibility for enforcing the quota and other restrictions without any supervision by international observers. Steps are being taken to change this. It is to be deplored that scientific methods were not introduced sooner and that even now stricter enforcement is necessary, but recent restrictions by the Commission represent a major accomplishment in management of a world resource, one that will survive only if man and nations cooperate to save it.

PROBLEMS

1. Refer to Figure 1. Find the totals of the catch of blue whales for the three year groups of 1931, 1932, 1933 and 1955, 1956, 1957. What is the approximate percent reduction in the whale catches?

2. How many blue whales were caught in the 1954–55 season? The 1961–62 season? Does this change show that fewer whaling vessels were sent out in 1961–62 than in 1954–55? (Hint: use Figure 2).

3. Refer to Figure 2. Use the catch per day method to estimate the number of blue whales alive in 1959.

4. Explain the earplug method of determining the age of a whale.

5. A biologist sets up camp along the banks of a river. From his campsite, he throws out a net and catches 100 fish, which he marks. The next year he returns to the campsite and traps another 100 fish in his net. Twenty of these bear the biologist's mark. The biologist estimates that there are $(100 \div 20) \times 100 = 500$ fish in the river. Do you agree with this calculation? Explain your answer.

6. Suppose the whale catch in 1980 is 10,000. In 1981, the whaling fleet is doubled, but the catch per day remains the same as in 1980. How many whales do you estimate there were at the beginning of 1981? Explain your reasoning.

REFERENCES

D. G. Chapman, K. R. Allen, and S. J. Holt. 1964. "Reports of the Committee of Three Scientists on the Special Antarctic Investigations of the Antarctic Whale Stocks." *Fourteenth Report of the International Whaling Commission.* London. Pp. 32–106.

J. A. Gulland. 1966. "The Effect of Regulation on Antarctic Whale Catches." *Journal du Conseil,* 30:308–315.

N. A. Mackintosh. 1965. *The Stocks of Whales.* London: Fishing News.

Scott McVay. 1966. "The Last of the Great Whales." *Scientific American,* 215:2, pp. 13–21.

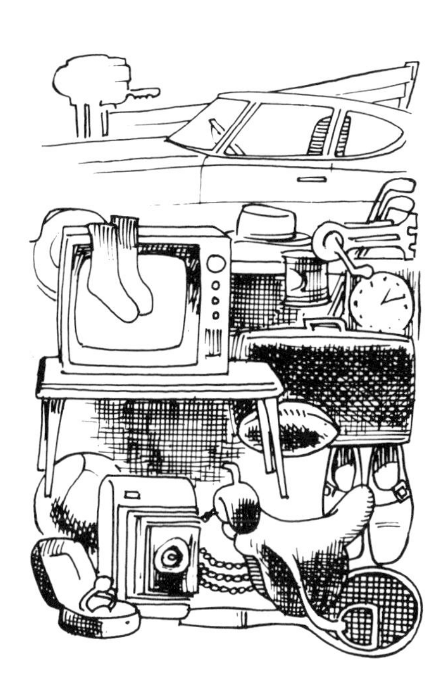

MAKING THINGS RIGHT

W. Edwards Deming *Consultant in Statistical Surveys*

WHAT IS THE STATISTICAL CONTROL OF QUALITY?

The statistical control of quality is the use of statistical methods in all stages of production—in design of product, in tests of product in the laboratory, in tests in service, for specifications and tests of incoming materials and assemblies, and for achieving economies in production, maintenance, and replacement of machinery and equipment, economies in inventory of parts for repairs of machinery, even economies in inventory to meet predicted demand.

Inspection is a very important function in production. The effects of instruments, machines, and human observations jointly create figures that must be transcribed onto forms constructed for the purpose. Faults recorded in inspection may be inherent to the product, or they may be caused by faulty instruments or gauges, or even by poor measuring practice.

We must be content in this article to limit ourselves to a few simple examples of statistical control of quality drawn from the production line. In the first two examples the aim will be to detect the existence of special causes of trouble, for the operator to correct. In the third example the aim will be to measure the effects of common (environmental) causes of trouble, for management to correct. In the real world, we are always working on both kinds of causes. We hope the reader will see in the examples the distinction between special causes and common causes and how they affect the variability of the process or lead to other kinds of trouble.

EXAMPLE 1: FUDGING THE DATA

Figure 1 shows the distribution of diameters in centimeters, these being the results of the inspection of 500 steel rods. Such a graphic representation of a distribution is called a histogram. The lower specification limit (abbreviated LSL) of the diameter of these rods was 1 centimeter. Rods smaller than 1 cm. would be too loose in their bearings, and such rods would be thrown out (rejected) in a later operation, when they must be fitted to a hole. Rejection means loss of all the labor that was expended on the rod up to this point, as well as loss of material and of overhead expense.

The horizontal axis in Figure 1 shows the centers of intervals of measurements; for example, 0.998 stands for rods that measured between 0.9975 and 0.9985 cm. The vertical axis is labeled to show the number of rods that fell into an interval of 0.001 cm. on the horizontal axis. For example, about 30 rods were in the interval centered at 0.998. It appears from the distribution that $10 + 30 + 0 = 40$ rods failed because they were too small.

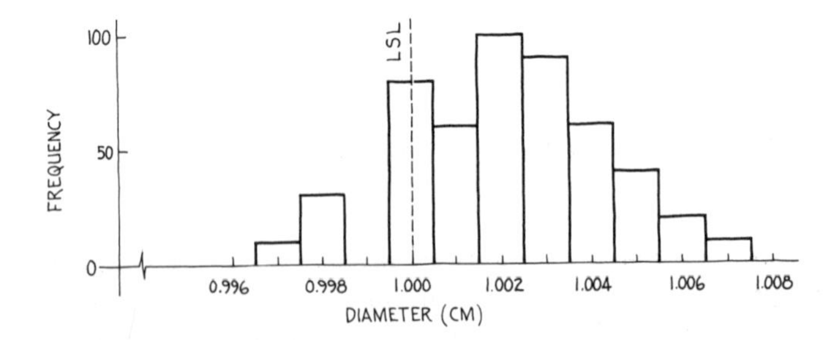

FIGURE 1

Distribution of measurements on the inside diameters of 500 steel rods. The chart detected the existence of a special cause of variation, a fault in recording results of inspection

A distribution is one of the most important statistical tools, when used with skill, yet it is extremely simple to construct and to understand.

Figure 1 is trying to tell us something. The peak at just 1 cm. with a gap at 0.999 seems strange. It looks as if the inspectors were passing parts that were barely below the lower specification, recording them in the interval centered at 1.000. When the inspectors were asked about this possibility, they readily admitted that they were passing parts that were barely defective. They were unaware of the importance of their job, and unaware of the trouble that an undersized diameter would cause later on.

This simple chart thus detected a special cause of trouble. The inspectors themselves could correct the fault. When the inspectors in the future recorded their results more faithfully, the gap at 0.999 filled up. The number of defective rods turned out to be much bigger, 105 in the next 500, instead of the false figure of $10 + 30 + 0 = 40$ in Figure 1.

The results of inspection, when corrected, led to recognition of a fundamental fault in production; the setting of the machine was wrong. It was producing an inordinate number of rods of diameter below the lower specification limit. When the setting was corrected and the inspection carried out properly, most of the trouble disappeared.

The upper specification limit had its problems also, but they were not so serious. A rod that is too large in diameter can be tooled off to fit. This is not the economic way to achieve the right dimension, but it is cheaper than to lose all the labor expended up to that time on the rod. The next problem was accordingly to increase uniformity and work on the correct centering of the average diameter, to reduce the number of defectives with wrong diameters.

EXAMPLE 2: DETECTING A TREND

The second example deals with a test of coil springs one after another as they come off the production line. These springs are used in cameras of a certain type. According to the specifications, the spring should lengthen by 0.001 cm. for each gram of pull. These springs are relatively expensive, and are supposedly made to exacting requirements. The length of any horizontal bar in the histogram at the right in Figure 2 shows how many springs the inspectors recorded with the elongation shown. We have turned it sidewise for convenience. This histogram represents measurements on 50 springs manufactured in succession. It will be noted that the distribution is symmetrical and is centered close to the specification; furthermore all 50 springs were within the upper and lower specification limits. One might be tempted to conclude from this histogram alone that the production of this spring presents no problems. However, another simple but powerful statistical tool, called a *run chart,* indicates trouble, as we now explain.

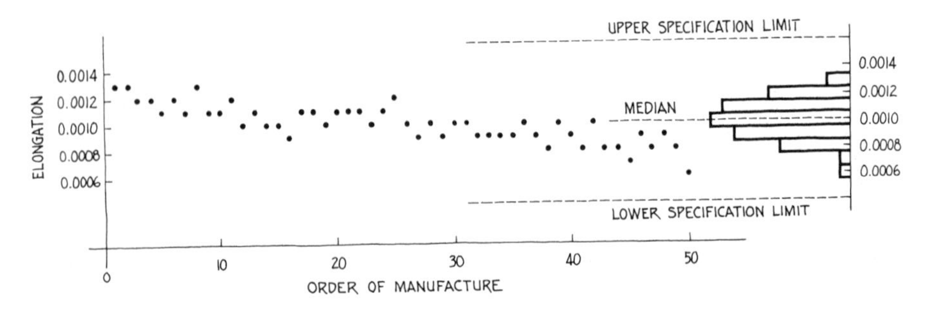

FIGURE 2

Run chart for 50 springs tested in order of manufacture. The chart shows a definite trend downward and thus reveals the existence of a special cause of variation, which it is important to correct. The frequency distribution alone could not detect this trouble

A run chart is merely a running record of the results of inspection. The horizontal scale shows the order of the item as produced, and the vertical axis shows the measurement for that item. In Figure 2, the elongations of the 50 successive springs are plotted on the vertical scale. A run chart has several simple uses. For example,

(1) A run of six or seven consecutive points lying all above or all below the median—the middle point in height—signifies with near certainty the existence of a special cause of variation, usually a trend.
(2) A run of six or seven points successively progressing upward or successively progressing downward has the same significance.

In no instance in Figure 2 is there a run up or a run down of length 6. It so happens that the median of the 50 points falls midway between the upper and lower specification limits. This would be good, but we note that the opening burst of points at the left of the figure has 10 points in succession above the median. Fifteen out of 18 points after point 29 fall below the median. These observations give a statistical foundation for the conclusion that, although the points vary up and down, there is a general drift downward. You may feel that your eye was good enough to detect this trend without knowing from theory that a run must have six or seven points above or below the median to detect with near certainty the existence of trouble, and in this example, you would be correct, but in more complicated examples such trends are often not detectable by eye.

Knowledge of what lengths of runs are required to indicate trouble is also valuable but secondary in problems of production. Indeed, it is an important statistical point that some of the most powerful statistical techniques are simple, as in our examples here. It was their widespread use, which began about 1942, that laid the foundation for the statistical control of quality, which of course has since grown into all phases of management. This movement led to the organization years ago of the American Society for Quality Control, over 23,000 strong in 1970.

In our camera-spring example, either the production process is in trouble or the apparatus used for testing is giving false readings. Correction is vital, whatever be the source of the trend. If it is the tension of the spring that is drifting downward (and not the testing apparatus), defective springs will be produced in the immediate future. If the source of the trend is faulty testing, then the tests are misleading, and may have been giving faulty reports on all the springs produced recently.

In this particular case, the trouble lay in a thermocouple that permitted the temperature to drift during the annealing of the springs. The process was headed for trouble. The simple run chart detected the trend before trouble occurred. The operator himself, seeing the trend, was able to head off trouble.

The reader may note that the histogram and the run chart in Figure 2 were plotted from the same data, yet they tell different stories. The histogram by itself gives no indication of anything wrong; it could have indicated unsatisfactory positioning. The run chart, however, leads us to suspect the existence of something wrong, a trend that, unless corrected, would soon lead to the production of defective springs.

It is interesting to note that if the points in Figure 2 had been plotted in random order instead of one after another in the order of production (1, 2, 3, and onward to 50), the run chart would have lost its power to detect a trend. Statisticians are thus not only concerned with figures, but with the relevant figures. In this instance, the order of production was relevant—very relevant—and was used to make the run chart. The histograms in Figures 1 and 2, on the other hand, do not make use of the order of production. They would remain unchanged, regardless of order: they depend only on the numbers recorded as results of inspection. The histogram in Figure 1 nevertheless did its work; it told us that something was wrong (namely, in the inspection itself). A run chart in connection with Figure 1 would not have added any relevant information. The histogram in Figure 2, however, was helpless to detect the existence of anything wrong. Judging by it alone, without the run chart, we could not have detected impending trouble.

EXAMPLE 3: MEASUREMENT OF COMMON (ENVIRONMENTAL) CAUSES

The first two examples dealt with special causes, specific to a designated worker or to a machine or to a specific group of workers. Statistical techniques point to specific sources of trouble when the process is nonrandom. The same statistical methods also tell the worker to leave things alone, to avoid overadjusting when attempts at adjustments would be ineffective or cause even greater variation than now exists.

There is another kind of problem that faces the management of any concern. No matter how skilled the workers, and no matter how conscientious, there will be at least a bedrock minimum amount of trouble in production owing to common or environmental causes. All the workers in a section work under certain conditions fixed by the management, or one might say, by the environment, which only the management can alter. For example, all the workers use the same type of machine or instrument. They are all doing about the same thing, and are using the same raw materials (which might be semifinished assemblies). They must put up with the same amount of noise and smoke.

It used to be supposed by management that all troubles came from the workers: that if the workers would only carry out with care the prescribed motions (soldering a joint, placing a part, turning a screw), then the product would be right with no defectives. This kind of reasoning does not solve the problem. Alert management can look into the problem with "infrared vision," supplied by statistical techniques.

An example was a small factory that made men's shoes. The machinery that sews soles is expensive. Time that an operator spends rethreading the machine and adjusting the tension after a break in the thread is time lost. Minutes lost may add up to hours and even days in the course of a month. There is not only the loss of rent paid for the machine and wages of the operator, but loss of labor and materials, nonproductivity of floor space, light, and increases in general overhead expenses. In this factory, about 10% of the working time was being spent rethreading the machines and adjusting the tension. Management was rightly worried. The trouble became obvious with a bit of statistical thinking. Observations on the proportion of time lost by the individual workers provided data for a chart similar to Figure 3. This figure showed that all the operators were losing about the same amount of time rethreading their machines. In fact, the time lost per day per man showed a pattern of randomness. This uniformity across operators could only point to the environment. What was the trouble?

The trouble turned out to be the thread. The management of the company was trying to save money by buying second-grade thread that cost 10¢ per spool less than first-grade thread. Penny wise and pound foolish! The

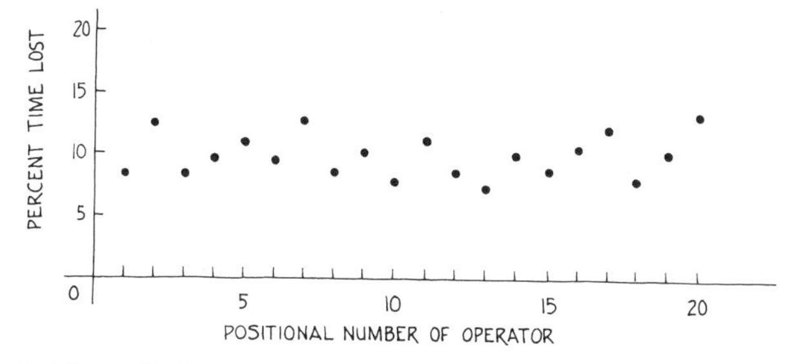

FIGURE 3
Time lost by each of 20 operators

savings on thread were being wiped out and overwhelmed many times over by troubles caused by poor thread.

A change to thread of first grade eliminated 90% of the time lost in rethreading the machines, with savings many times the added cost of better thread.

What is the distinction between this example and examples 1 and 2? In examples 1 and 2, the workers themselves could make the necessary changes, and they did. In this example, the operators were helpless. They could not put in an order for thread of first grade and scrap the bad thread. Their jobs were rigid—work with the materials and machines supplied by the management. They all worked with the same bad thread; that is, they all worked under a common cause of trouble. Management is responsible for common (environmental) causes; therefore only management could change the thread.

But how is management to know that there are common causes of trouble? The answer is simple: common causes are always present. Management needs a better answer, however; management needs a graphical or numerical measure of the magnitude of the trouble wrought by common causes. Without statistical techniques, management can have no accurate idea about the magnitude of the trouble being caused by conditions that only management can change.

Charts such as Figure 3 tell the management that there is a problem, that the time lost on rethreading will not go below 10% until management makes some fundamental change. The change in thread in example 3 was a fundamental change. What to change is not always as easy to perceive as it was in this example, however. Sometimes a series of experiments is required to discover the main causes of trouble. Statistically designed experiments have led to the identification of common causes such as raw materials not suited to the requirements, poor instruction and poor supervision (almost synonymous with unfortunate working relationships between foremen and production workers), and vibration.

Shift of management's emphasis from quantity to quality is one common environmental cause of trouble. The production workers continue to work with emphasis on quantity, not quality. Discussion of methods by which management may direct the shift from quantity to quality, however important, is beyond the scope of this essay.

PROBLEMS

1. In Example 1, why is having too large a diameter rod less serious than having too small a diameter?

2. While conducting a large-scale community health audit, a medical researcher developed a fast method of measuring blood pressure. A statistician claimed to have discovered a flaw in the method just by looking at the following histogram of 1000 blood pressures (as measured by the new method).

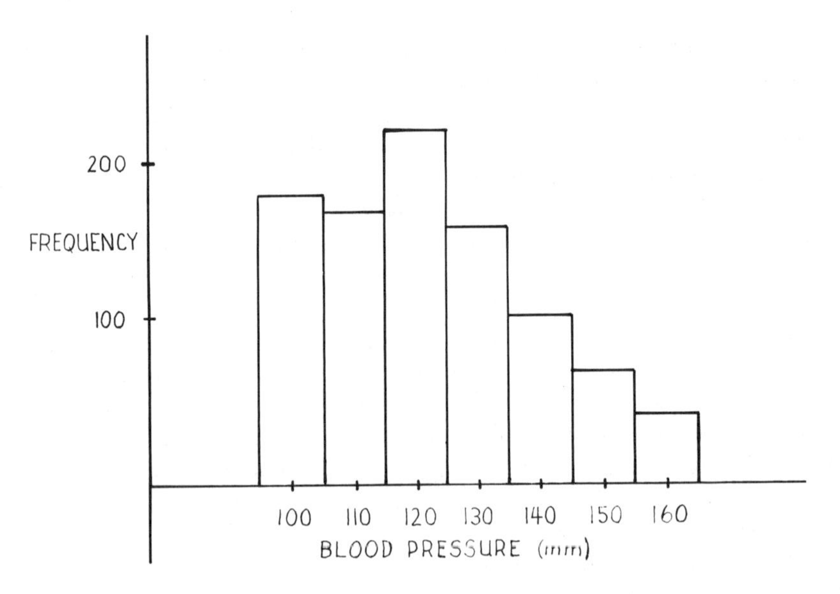

What seems to be wrong with the method?

3. Suppose the thermocouple problem had not existed for the camera spring example. Do you think the histogram of 50 spring pulls would be as spread out, more spread out or less spread out than the one in Figure 2?

4. In automobile manufacture, it has often been found that cars produced on Mondays and Fridays are more frequently defective than those on Tuesdays, Wednesdays and Thursdays. Why do you think this is so? If you were part of the management, how might you attempt to correct this?

5. When is it desirable to use a run chart for a manufacturing process? Why? Justify your answer.

6. Referring to Figure 3, which worker lost the most time rethreading? How much time? Which lost the least? How much time?

7. In a paper-cup factory, the cups are conveyed along a belt with four operating positions: the first operator cuts and glues the cups, the second coats them with plastic, the third counts and packages them, and the fourth places the packages in large cartons and seals the cartons. Management is concerned that too many cups are ruined during production. The quality control department studied the site at which cups were ruined during a shift (there were 8 belts in operation).

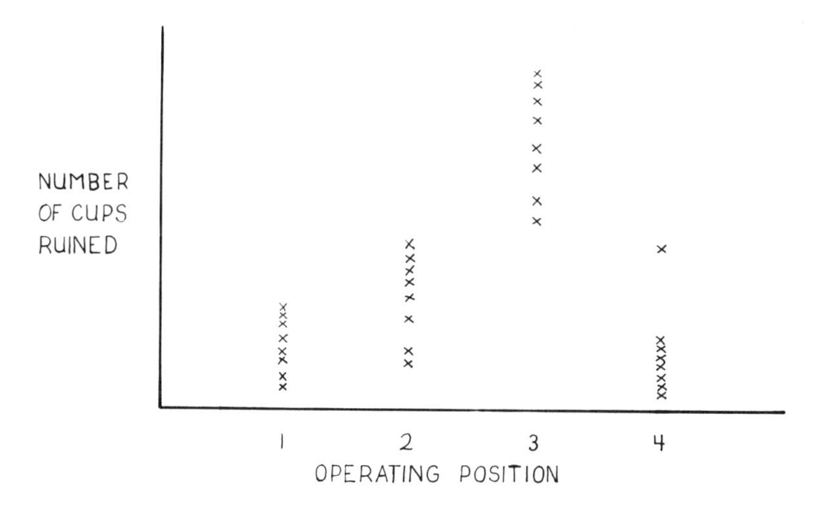

Is this a problem which requires a solution at the level of management?

8. In the 7 to 4 shift at the paper-cup factory, the operators of the plastic-coating machines get a break from 9–9:30 and a lunch break from 1–1:30. During these times their machines are idle. Management is worried because too many cups are being produced with an unacceptable amount of plastic coating, some with too much and some with too little. The quality control department prepared charts for each machine, all of which had the following appearance (each dot stands for a batch of cups).

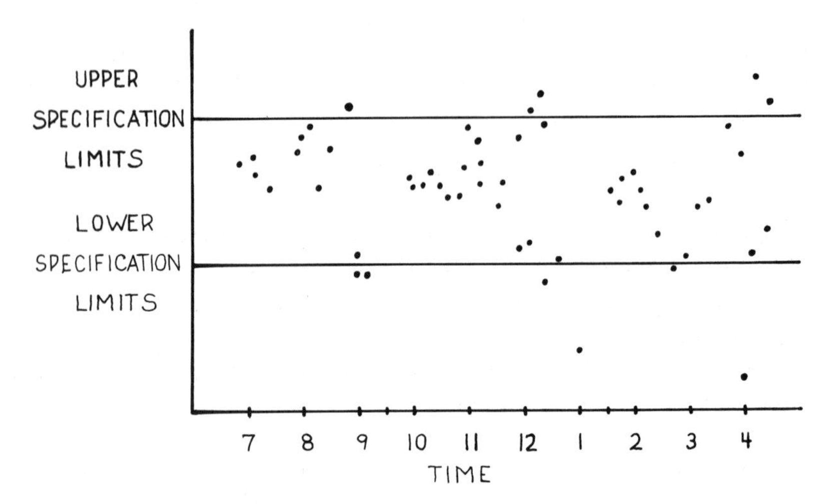

One manager suggested that the plant might save money by giving the operators two extra coffee breaks. Why?

DRUG SCREENING: The Never-Ending Search for New and Better Drugs

Charles W. Dunnett *Lederle Laboratories Division, American Cyanamid Company*

PROBABLY EVERYONE can recall reading about a small boat lost at sea or an aircraft down in some unpopulated area involving a search for survivors. If the search is successful, both rescuers and rescued receive wide public acclaim. On the other hand, if the search is unsuccessful, the episode is soon forgotten. In neither case do we see much in the press about the planning and organization of the search. We can imagine, however, what a tremendous effort it must be to map out the area to be explored, to marshal the needed resources in aircraft and other vehicles, and to use all the available people and equipment so as to maximize the chance of a successful outcome in the shortest possible time.

Pharmaceutical companies conduct somewhat similar searches for new drugs, but their search is continuous. Just as a rescue team has some idea where to look, perhaps based on a radio message received from the victims, so research chemists often know what types of chemical structures to look for to treat a particular disease, and the chemists can set about synthesizing compounds of the desired type. Sometimes, however, their knowledge may be vague, resulting in such a wide range of possibilities that many, many compounds have to be made and tested. In such a case, the search is very lengthy and requires years of effort by many people to develop a useful new drug.

Researching thousands of compounds for the few that might be effective requires highly organized, efficient testing methods. With an inefficient procedure, progress will be slow and the testing laboratory may never get to the "good" compounds. Of course, the procedure used to test each compound must have a high degree of accuracy and must be capable of detecting a good compound.

Testing a long series of compounds in the search for any that have useful biological properties is known as *drug screening*. Drug screening requires laboratories, technical personnel to operate them, and appropriate apparatus and instruments. Ordinarily, laboratory tests are conducted with animals, which must be housed, cared for, and observed in the laboratory. Space limitations together with the limitations of the staff severely restrict the number of compounds that can be assessed. A single laboratory cannot hope to test more than several hundred, or, perhaps, a few thousand, compounds annually for a specified biological activity. This testing rate cannot even keep up with the rate of synthesis of new compounds. Yet, if all the laboratories throughout the entire pharmaceutical industry are considered, much is accomplished. It is estimated (Arnow 1970), for example, that approximately 175,000 substances are subjected to biologic evaluation each year.

The screening test is only the first hurdle along the path of developing an effective drug; much further testing and investigation needs to be done before the drug can be tried on humans. In fact, only about 20 of the 175,000 substances tested finally become available in the local drug store.

Anything that improves the efficiency of the testing procedure increases the chance of discovering a new cure. Naturally, the biologist designing a new screening procedure attempts to use an animal and test system that reflect the human disease for which a cure is being sought as accurately as possible and yet are easy to use in the laboratory. We will not deal with this aspect of the problem here, but rather with the contribution of statistics to improvement of drug screening procedures.

AN EXAMPLE: ANTICANCER SCREENING

In drug screening, the animals of a sample treated with a particular compound are observed to determine whether the treatment is having a desirable effect. Perhaps all that is noted in each animal is whether or not it is "cured" or shows improvement. The result of the test can be expressed simply as the number of cured or improved animals out of the total number tested. Since not every animal responds to treatment in exactly the same way, the number cured might be 0, 1, 2, or all the way up to the whole sample size.

More often, a measurement of some sort is made upon each animal, the magnitude being indicative of the treatment effect. For example, in anti-cancer screening, after implanting cancer cells in mice, the investigator treats them with the test chemical compound to see whether it retards the growth of the cancer tumors. After treating the mice for a fixed length of time, he removes the tumors and weighs them. For comparison, a similar group of untreated "control" animals is handled in the same way except for the absence of the chemical treatment. If the chemical is effective, the tumor weights of the treated mice should be less than the tumor weights of the control mice. The statistician's job is to help decide, on the basis of the numerical values obtained, whether the tested compound merits further investigation.

The following are actual tumor weights observed in three animals treated with a test compound and in six untreated control animals:

Treated: 0.96, 1.59, 1.14 grams
Controls: 1.29, 1.60, 2.27, 1.31, 1.88, 2.21 grams

The reason for having more control animals than treated is that, in one experiment perhaps 30 to 40 different compounds will be tested, each one in a different set of three animals. One control group suffices, however, to compare with the results from all the test compounds; hence it is desirable to have it larger in size in order to obtain a more precise control average.

Note the high variability from one animal to another. This is typical of the biological variation in screening tests that makes it difficult to determine with certainty whether the treated animals have really improved as a result of the treatment. How might a statistician go about deciding whether a compound has any merit?

DRUG SCREENING AS A DECISION PROBLEM

In drug screening, two actions are possible: (1) to "reject" the drug, meaning to conclude that the tested drug has little or no effect, in which case it will be set aside and a new drug selected for screening, and (2) to "accept" the drug provisionally, in which case it will be subjected to further, more refined experimentation.

To abandon a drug when in fact it is a useful one (a *false negative*) is clearly undesirable, yet there is always some risk of that. On the other hand, to go ahead with further, more expensive testing of a drug that is in fact useless (a *false positive*) wastes time and money that could have been spent on testing other compounds.

Thus, we are faced with what is known in statistics as a *decision problem:* how to use available experimental data to choose between alternative courses of action in a rational way.

AN APPROACH TO SOLVING THE PROBLEM

What would the drug screening investigator like to achieve? A virtually unlimited supply of compounds is available for testing, more than he can hope to test over any reasonable period of time. Most of them lack the biological activity he is searching for, but (hopefully) a few of them possess it. His goal is to find as many of these active compounds as possible with the facilities at his disposal.

As a hypothetical, but not entirely unrealistic, example, let us suppose there are 10,000 compounds of which 40 are active and the remaining 9960 are inactive. The investigator is screening the drugs for antitumor activity, and he tests them by treating groups of three tumor-bearing mice with each compound, comparing the resulting tumor weights with the corresponding values for six control animals observed in the same experiment. He wishes to accept or reject each test compound on the basis of the observed tumor weights. Typically, about 50 compounds per week can be tested in this way, so that about four years' work will be required for all 10,000 compounds. Over this period of time, it is inevitable that some of the 9960 inactive compounds submitted to the screening will pass. The follow-up tests required on these false positives will require test facilities that could have been employed to screen some new compounds. Often the next step after a compound passes the screening step is to carry out a "dose-response" study, which consists of treating groups of animals with several dose levels of a drug in order to determine the relationship between dose and response. At this step, the inactive false positive compounds generally will be eliminated. Suppose, for illustration, that 30 animals are required for each compound at this step. This means that to follow up each compound accepted by the screening, it will be necessary to forgo or postpone the testing of ten new compounds.

Consider the experimental data given above. The mean, or average, tumor weight for the treated animals is $(0.96 + 1.59 + 1.14)/3 = 1.23$ grams. Comparing this with the corresponding mean for the six control animals, $(1.29 + 1.60 + 2.27 + 1.31 + 1.88 + 2.21)/6 = 1.76$ grams, we see that a reduction of $1.76 - 1.23 = 0.53$ gram has apparently been obtained. If the drug has no effect, a zero reduction would be "expected," but, of course, the variability of the animals makes it likely that some difference between the two means would occur even if the drug has no effect Thus, the researcher must decide how large a difference he requires the drug to show before he decides to "accept" it.

Let us assume for the moment that a reduction of 0.53 gram in tumor weight is not enough to convince our investigator that the compound is an active one; let's say that he requires a reduction of 0.70 gram before he will permit the compound to pass the screening test. What are the consequences of setting a cut-off value of 0.70 gram in screening the 10,000 compounds?

What he really needs to know is how many of the 9960 inactive compounds and 40 active compounds will pass the test. He could go ahead and screen the compounds using the 0.70-gram criterion, but it would take several years to collect the results: this would be rather late to find out that he made an unwise choice in the cut-off criterion.

Fortunately, there is another way to get at this problem. Statisticians have a unit, or yardstick, called the *standard error*, which they use to determine how often measures like the difference between two means will exceed any specified limit. An estimate of the standard error needed in our case could be calculated from the observed data, but it would be unreliable because of the small number of observations. It is possible, however, to estimate the standard error accurately from other data of the same type, which, in routine screening, are available in large quantities from past records. Suppose that, for tumor weights, the standard error of a difference between a mean of six control tumor weights and a mean of three treated tumor weights is known to be 0.35 gram.

The next step is to divide the difference between the cut-off value and the *expected* value of the test statistic by this standard error. For an inactive compound, our researcher would expect the two mean tumor weights to be the same; hence the test statistic is expected to have the value zero. The cut-off value of 0.70 gram, therefore, is $(0.70 - 0)/0.35 = 2.0$ standard errors from the zero expectation. Consulting a table of the *normal distribution* (available in most statistics texts) he finds that a deviation of 2.0 or more standard errors from the expected value occurs with a probability of .0228. This means that he can expect to observe a misleading reduction in tumor weight exceeding 0.7 gram for approximately 23 out of 1000 inactive drugs submitted to screening. In other words, of the 9960 inactive compounds, $.0228 \times 9960 = 227$ of them can be expected to pass as false positives.

Consider next the 40 active compounds: how many of them can be expected to pass? This is a more difficult question, because the answer depends on how active they really are. It is not to be expected that even a very active compound will eliminate the tumor completely in the relatively short time the animals are under treatment. Modest decreases in tumor size are the most that researchers can hope for.

Let us assume for now, rather arbitrarily, that the 40 active compounds are each capable of reducing tumor weight by 0.7 gram (the same as that required by our cut-off criterion). As each active compound is tested, some actually will result in a reduction of more than 0.7 gram, while others will give rise to smaller reductions, because of the inevitable variability of the animals. Therefore, only half of the actives can be expected to show a reduction exceeding the stipulated cut-off value and, hence, to pass the test.

This means that on the average, the accepted compounds will consist of 20 actives (true positives) and 227 false positives. The follow-up testing required on each is the equivalent of ten screening tests, so, in addition to the 10,000 screening tests, there will be $247 \times 10 = 2470$ follow-up tests, or 12,470 tests in all. For this effort, the researcher can expect to find 20 active compounds under the assumptions about the composition of the original set of 10,000 compounds and the degree of activity of the actives. It is useful to express this as a "yield" per thousand tests. Using 0.7 gram as the cut-off value, the resulting yield is

$$\frac{20}{12{,}470} \times 1000 = 1.60 \text{ actives/thousand tests.}$$

Next, consider the consequences of altering the criterion for a compound to pass the screen. Suppose that instead of a 0.7 gram reduction in tumor weight, a cut-off value of 0.6 gram is used. This corresponds to a deviation of $0.6/0.35 = 1.71$ on the normal curve, and the corresponding probability obtained from tables of the normal distribution is .0436. Thus, of the inactive compounds tested, $.0436 \times 9960 = 434$ false positives are expected to occur.

What about the 40 active compounds? The researcher still assumes that each has a degree of activity capable of producing a reduction of 0.7 gram on the average. How many of them will actually produce a reduction of 0.6 gram or more? Expressing the difference of 0.1 gram as a deviation on the normal curve by dividing by the standard error, he obtains $0.1/0.35 = 0.29$; from the normal tables, he finds that the corresponding probability is .6140. Hence, using 0.6 gram as the passing criterion, he can expect $.6140 \times 40 = 25$ active compounds to pass. This results in $434 + 25 = 459$ compounds passing the screen, so the total testing effort is now $10{,}000 + 459 \times 10 = 14{,}590$ tests. Thus the yield per thousand tests is

$$\frac{25}{14{,}590} \times 1000 = 1.71 \text{ actives/thousand tests,}$$

an increase of 0.11 over the yield obtained using 0.7 gram as the cut-off value.

Now it should be clear how to go about "optimizing" the choice of the cut-off value for the given hypothetical structure of the compounds. It is necessary to try various cut-off values to determine the yield for each in the above way. Figure 1 shows a curve of the computed yield plotted against

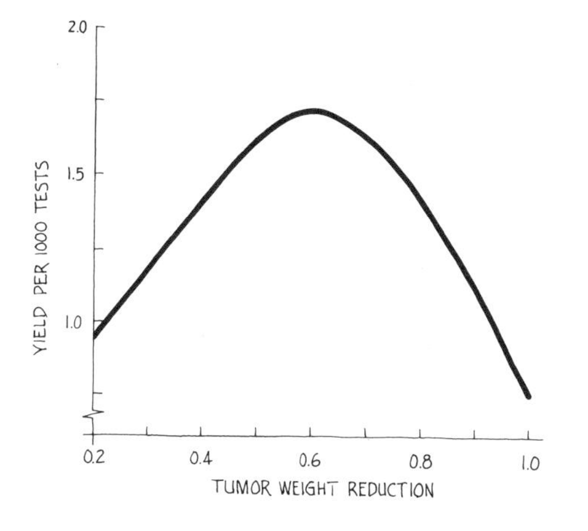

FIGURE 1
Yields for one-stage screening tests

the corresponding cut-off value. It can be seen that the point of maximum yield is easily determined and, in fact, occurs very close to a cut-off value of 0.6 gram reduction. Of course, this applies only to the particular hypothetical mix of active and inactive compounds that we have assumed; we will come back to this point later. For this particular mix, a yield of 1.71 per thousand tests appears to be about the best that can be obtained.

SEQUENTIAL PROCEDURES FOR DRUG SCREENING

It is possible, however, to do still better by using another type of screening. Statistical theory tells us that a kind of test procedure, known as a *sequential test,* is more efficient for reaching decisions of the type we need. In sequential testing, the number of tests is not fixed in advance; rather, the testing proceeds sequentially and the number of tests depends upon the results observed. This gives us the option, after observing a test result on a compound, to forgo making an immediate decision about that compound and to subject the compound to another test. From the results of a second test averaged with those of a first test, we may conclude that further testing is unnecessary, or we may forgo a decision again and wait for a third test result on the compound. This process could go on until a decision is reached to accept or reject the compound. Of course, if too many stages are permitted, the procedure loses efficiency because of the delay and because of the extra trouble involved in keeping a sufficient supply of the compound on hand to repeat the test. In drug screening applications, it has usually been found best to limit the testing to two or three stages.

A sequential procedure is more efficient because it allows us to compile additional data on the compounds about which doubt exists, without much increase in the average amount of testing required, because additional testing needs to be done only on a relatively few compounds.

Suppose that we decide to reject a compound without further testing if a tumor weight reduction of less than 0.2 gram is observed on the first test. If, instead, a reduction of more than 0.2 gram is obtained, we will test the compound again on another group of mice. At the end of this second stage of testing, the tumor weight reduction will be measured as usual, and its value will be averaged with the value obtained for that compound in the first stage. If the average of these two values exceeds 0.5 gram, we will accept the compound; otherwise it will be discarded. For obvious reasons, a procedure of this type is called a *two-stage test* (the procedure discussed in the previous section is called a *one-stage test*).

For illustration, consider our numerical example. For that set of data, the observed tumor weight reduction turned out to be 0.53 gram. Using a one-stage test, we found that the optimum cut-off value was 0.6 gram, which would require this compound to be discarded. With the two-stage test described in the preceding paragraph, however, it would be tested again. Suppose another test (using three treated and six control animals) gave a tumor weight reduction of 0.49 gram. Averaging the two results, 0.53 and 0.49, we obtain 0.51 gram; this is enough for the compound to be accepted.

What are the consequences of using the two-stage test on the hypothetical mix of 10,000 active and inactive compounds? First of all, the amount of testing obviously will be increased because, for some compounds, two tests will have to be run. The amount of increase can be computed. For an inactive compound, the probability of obtaining a tumor weight reduction in excess of 0.2 gram turns out to be .284. Thus, $.284 \times 9960 = 2829$ of the inactive compounds will require two screening tests to permit a decision, the remainder requiring only one test. For an active compound, the corresponding probability is .923; hence, $.923 \times 40 = 37$ of the actives will be tested a second time. Therefore, the total number of screening tests required to screen the 10,000 compounds is $10{,}000 + 2829 + 37 = 12{,}866$ (compared to 10,000 in a one-stage procedure).

It is also fairly easy to work out the probability that an inactive compound passes the screening under the sequential procedure; this may be done for an active compound as well. For an active compound, the probability is .77; thus, $.77 \times 40 = 31$ active compounds can be expected to pass. For an inactive compound, the probability is .020; hence, $.020 \times 9960 = 199$ false positives will occur. Therefore, a total of 230 compounds are expected to pass, entailing an additional testing effort equal to 2300 tests. This makes altogether $12{,}866 + 2300 = 15{,}166$ tests. The yield per thousand tests for this scheme is

$$\frac{31}{15{,}166} \times 1000 = 2.04 \text{ actives/thousand tests;}$$

note that this is higher than the best yield that can be obtained with a one-stage test.

By trying various cut-off values for the two stages of the test, an optimum two-stage screening test can be devised. The optimum yield turns out to be 2.15 per thousand tests, achieved with a cut-off value of 0.4 gram reduction at the first stage and 0.5 gram reduction at the second stage. The increase in yield from 1.71 per thousand with the best one-stage test to 2.15 per thousand using the best two-stage test represents an improvement of 26% in the efficiency of the screening test. This is an important advance because it means that useful cures for diseases can be found more quickly.

Further improvements can be obtained by considering more than two stages of testing. We will not go into the details here, but an optimum three-stage test can be devised to yield 2.33 per thousand. More than three stages can be considered, but it is questionable whether the increases in efficiency that theoretically could be obtained are worth the extra complication. (Complicated bookkeeping in a large laboratory has its own high costs in both potential errors and manpower.)

EFFECT OF ASSUMPTION ABOUT THE COMPOSITION OF THE COMPOUNDS BEING SCREENED

In the forgoing we showed how the statistician can optimize the choice of the cut-off values that determine whether a given compound should be accepted, rejected, or tested again. The optimum, however, was based on a hypothetical mixture of active and inactive compounds being screened. We assumed that 10,000 compounds contained 9960 inactives and 40 actives; moreover, we assumed the active compounds had a degree of activity capable of reducing the animal tumor weights by 0.7 gram.

Of course, the statistician must repeat the whole process for other assumptions about the composition of the compounds being screened to determine the effects on the optimum cut-off values. It turns out that the actual *number* of actives assumed to be in the mix does not affect the screening procedure. In other words, if the 10,000 compounds contained 100 actives, or 4 actives, or only 1, instead of 40, the same screening criteria would produce an optimum yield. (Of course, the actual *magnitude* of the optimum yield would go up or down with the number of actives assumed to be present.) On the other hand, the degree of activity assumed for the active compounds does affect the procedure. All the statistician can do here is to work out the best procedure for various degrees of activity and let the screening investigator decide on the basis of his knowledge what level of activity of interest to him is most likely to occur. Then the statistician can tell him what screening criteria will produce the optimum results for that level of activity. It is also possible to consider a mixture of two, three, or even several different levels of activity among the active compounds.

OTHER VARIABLES IN THE SCREENING PROCEDURE

In the preceding discussion, it was assumed that the test procedure itself was fixed and the only variables were the cut-off criteria. Other factors, however, also can be altered, for example, the number of animals used in each test. Is the choice of three animals for each test compound and six animals for control really best? A reduction in either of these numbers would enable the investigator to test more compounds in each experiment. On the debit side, however, this would increase the standard error of the test statistic, which would have the undesirable effect of increasing the chances for a compound to be misclassified.

The statistician can study the effect of changes in these variables on the theoretical yield, by carrying out calculations similar to those we have already described. The object is to achieve, with the test facilities that are available, the greatest expected yield in terms of active compounds found. Of course, it is impossible to guarantee what results will actually be obtained. No matter how efficient a screening procedure is, in fact, there may be *no* active compounds presented for screening. In the end, we must depend upon the ingenuity of the chemists to produce compounds that have the desired activity.

APPLICATIONS

Statistical studies of the sort described here have improved the efficiency of many of the routine screening procedures in our laboratories. The anticancer screening program is discussed in detail by Vogel and Haynes (1962), who state that an increase in the screening rate from 450 compounds per year to 1300 per year has been achieved. I could provide a happy ending to this tale if I could tell you about the discovery of a new cure for cancer as a result, but although some possible leads have been discovered, it seems that we are still a very long way from this goal. Past successes in other areas, such as the remarkable discovery of the antibiotic Aureomycin after the screening of over 4000 soil samples, are convincing proof, however, that drug screening plays a necessary and important role in the never-ending search for new and better drugs. A recent book by Arnow (1970) contains an interesting account of this and other aspects of drug research.

The statistical aspects of drug screening are similar to screening and selection problems in other fields. For example, in the development of new and improved strains of an agricultural crop, such as wheat, each potential new strain must be planted and grown, and measurements of its yield and other indicators of performance must be taken. On the basis of the results, some strains are chosen to be planted again on a wider scale, and eventually, after many repetitions of the cycle, a new variety may emerge to replace current standard varieties. The plant breeder, like the drug screener, must determine how best to use his facilities for testing various candidates in order to maximize the chances of success. My 1968 article gives a more general discussion of the problems of screening and selection.

PROBLEMS

1. Referring to Figure 1, what would the yield per 1000 tests be if .8 gram were used as a cutoff value?

2. Refer to Figure 1. The yield per one thousand tests is greater for a 0.6 gram cutoff than for a 0.4 gram cutoff. Which cutoff do you think would yield a greater *number* of active compounds after all 10,000 compounds had been screened. Why?

3. Define the term two-stage test. How is a two-stage test different from a one-stage test?

4. Suppose that two different laboratories were to screen the same 10,000 compounds using the same screening procedure. Would they necessarily obtain the same number of active compounds? Would they necessarily obtain the same active compounds if they obtained the same number of compounds? Why or why not?

5. Suppose by mistake the same drug was subjected to two different drug screening trials to determine anticancer activity. Is it possible for one trial to indicate that the drug has an effect in reducing tumor size, while the other trial indicates it has no effect? Explain your answer.

6. Why use more animals in the control than the treatment group in a drug screening experiment?

7. If the "dose response" study is more effective at eliminating useless drugs and identifying helpful ones than the screening trials described in the article, why not just skip the screening trials and subject each drug to a "dose response" study?

8. In the 2-stage sequential test described in the article, 12,866 screening trials were required, more than the 10,000 trials required for the 1-stage procedure. How then can the sequential test have a higher yield than the 1-stage procedure?

REFERENCES

L. E. Arnow. 1970. *Health in a Bottle; Searching for the Drugs that Help.* Philadelphia: Lippincott.

C. W. Dunnett. 1968. "Screening and Selection." D. L. Sills, ed., *International Encyclopedia of the Social Sciences* vol. 14. New York: Macmillan and Free Press.

A. W. Vogel and J. D. Haynes. 1962. "Experiences with Sequential Screening for Anticancer Agents." *Cancer Chemotherapy Reports* 22:23–30.

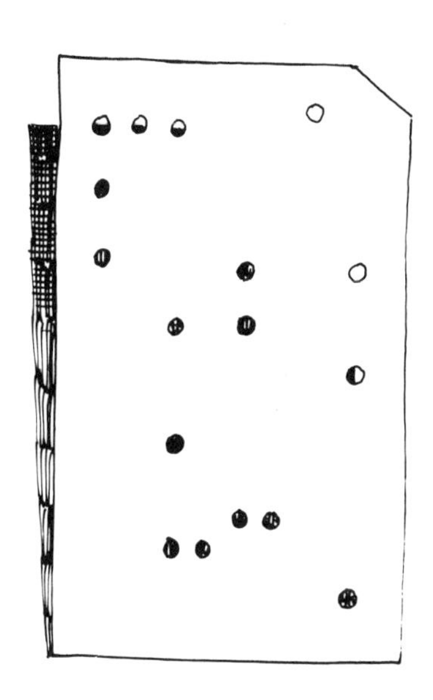

HOW TO COUNT BETTER: Using Statistics to Improve the Census

Morris H. Hansen *Westat Research, Inc.*

THE IMPORTANCE OF CENSUS RESULTS

In 1790, Thomas Jefferson gave to George Washington, our first President, the results of the first census of the United States of America. Every ten years since then, as provided in the Constitution, the decennial census has determined for the nation essential information about its people.

The basic constitutional purpose of the census is, of course, to apportion the membership of the House of Representatives among the states. From the beginning, the census has had many important purposes beyond the constitutional one. The development of legislative programs to improve health and education, alleviate poverty, augment transportation, and so on, is guided by census results. They are used also for program planning, execution, and evaluation. Now the distribution of billions of dollars a year from the Federal Government to the states, and from the states to local units, is based squarely upon census counts. (For example, if New York City is underenumerated by 2%, the result will be a loss of about 150,000 people and thus a loss of about one and a half million dollars a year in funds, or about $15,000,000 until the next census is taken.) Private business uses the census for many purposes, for example, plant location and marketing. Much social and economic research would be essentially impossible without census information.

The importance of the census to current problems such as poverty, health, education, civil rights, and others has brought about requests for shifting to a five-year, rather than a ten-year, census in order to keep information more nearly up to date. These requests, which are now under congressional review, have come from governors of states, mayors of large cities, scientists, businessmen, and many others.

THE JOB OF PLANNING AND TAKING THE CENSUS

The need for a fast and reasonably accurate census is fairly obvious. What may not be so apparent are the major problems involved in taking a census and making the results available promptly, and in adapting the census questions to serve current needs. In each census, some questions have been changed in response to new needs, but certain basic information has been consistently required. Questions in the 1970 population census included age, sex, race, marital status, family relationship, education, school enrollment, employment and unemployment, occupation and industry, migration, country of origin, income, and other subjects. In conjunction, a census of housing, with additional questions, was taken at the same time.

The job of organizing and taking the census is a major administrative and technological undertaking. Even though most of the questionnaires were filled out by the respondents themselves in a "census by mail" and the application of modern computers and other advanced technology has eliminated a lot of paperwork, the taking of a census requires the recruiting and training of about 150,000 people, most of them for only a few weeks of work. Once specific goals are set in terms of questionnaire content and desired statistical results, the massive job of organization and administration begins.

The system for canvassing and for collecting, receiving, processing, and summarizing the vast numbers of completed questionnaires must be planned. Specialized electronic and paper-handling machines designed and built at the Census Bureau automatically read the information recorded by respondents or enumerators. These complex machines first photograph on microfilm and then scan and read microfilm copies of the census forms that have been filled out mostly by the respondents themselves. The magnitude of the job is difficult to comprehend. For the 1970 census, approximately a quarter of a billion pages (counting each side of a relatively large sheet as a page) were handled. The results were recorded on magnetic tape, and then computers examined these results. In the process, the forms were edited for certain types of incompleteness or inconsistency, and adjustments were made automatically or special problem areas were identified for further manual investigation. The later steps of tabulation and printing for publication also were accomplished on electronic computers.

The approach to the job of taking a census differs from that of designing a totally new system, in that the census has been taken many times before, and the background and experience of the past serves to guide current efforts.

The availability of extensive past experience, however, has a disadvantage. There may be long traditions that have come to be regarded as essential, but that, in fact, only represent ways in which the job has been done in the past. For example, the tradition of taking the census by an enumerator canvassing an area and personally asking the questions of any responsible member of the household had been long established. This approach was regarded as proven by long use to be the only reasonable way to elicit information. Because the concepts in some census questions are difficult, it was thought that only a trained enumerator could ask the questions and elicit the proper information. But this view did not recognize the difficulties in training and controlling an army of temporary interviewers. Nor did it recognize that the responses obtained in the census interview situations were sometimes based on a misunderstanding of the questions or were spontaneous without the opportunity for a considered reply. Furthermore, the interviewer himself and his conceptions or misconceptions could importantly influence the response.

In the nineteenth century, many potential advantages of prior experience were lost, for the Census Bureau was not created as a permanent and continuing agency until 1902. It then became far more feasible, with a continuing staff, to benefit from lessons and experience of prior censuses. The situation for the 1900 and earlier censuses is illustrated in the following quotation from *The History of Statistics* (Cummings, 1918):

> Mr. Porter gives the following account of his experience, which must have been essentially that of every Superintendent of the Census.
>
> The Superintendent in both the last two censuses [1880 and 1890] was appointed in April of the year preceding the enumeration, but when I was appointed I had nothing but one clerk and a messenger, and a desk with some white paper on it. I sent over to the Patent Office building to find out all I could get of the remnants of ten years ago, and we got some old books and schedules and such things as we could dig out . . . I was not able to get more than three of the old men from this city . . . . I knew most of the old census people. Some of them were dead and some in private business . . . . But little over sixty days were allowed for the printing of 20,000,000 schedules and their distribution, accompanied by printed instructions to the 50,000 enumerators all over the country, many of them remote from railroads or telegraph lines . . . . Now

> to guide us in getting up these blanks, we had only a few scrapbooks that someone had had the forethought to use in saving some of the forms of blanks in the last census. He had taken them home, a few copies at a time, and put them into scrapbooks. The government had taken no care of these things in 1885, when the office was closed up. Some of them had been sold for waste paper, others had been burned, and others lost.

In addition to showing the potential gains from continuity and learning from the past this quotation suggests the great complexity of the censuses in the latter half of the nineteenth century. At that time, there was little in the way of a continuing statistical program in the federal government and as a consequence the decennial census was loaded with a range of questions that proved difficult if not impossible to collect through decennial census inquiries—hence the great number and variety of questions and forms. Many of these types of information are now collected through sample surveys or compilations from administrative records.

THE USE OF STATISTICS IN PLANNING AND TAKING THE CENSUS

Statistical concepts and methods have provided fundamental improvements to the census over the past 30 years. We might say that these improvements form a technological explosion. Part of this explosive change has come from introduction of the large-scale electronic computer, but even more of the change has come from statistical advances and the application of statistical studies. (In fact, development of the high-speed computer and modern statistical methods were both substantially motivated by census problems, and were in part carried out by census scientists or with census support.)

The Introduction of Sampling as a Tool for Census-Taking. Sampling was first used in collecting census information in the 1940 census. A series of questions was added for a 5% sample of the population. [Roughly speaking, this meant asking every twentieth person the additional questions (Waksberg and Pearl 1965).] This was a major advance, as the tradition for a century and a half had been universal coverage for every question. The questions asked of the 1940 sample included one on wage and salary income—income had not previously been a census question—one on usual occupation (as distinguished from current occupation; the 1940 census was planned during the Depression, when unemployment was very high and there was frequently a difference between a person's usual occupation and the occupation at which he was currently working), and several other questions.

The art and science of modern statistical sampling were evolving at the time of the 1940 census, and at the same time there was increasing public acceptance of sampling. It was possible to proceed with greater knowledge and confidence about what a sample would produce than would earlier have been possible. Thus, for a particular size and design of sample, statisticians could establish a reasonable range for the difference between a sample result and that obtained from a complete census. Suppose, for example, that a city had a total population of 100,000 persons of whom 30,000 were employed and earned wages or salaries, and suppose that 10,000 of these received wages and salaries of less than $2000 in 1939 (the year preceding the census). There would be approximately 5000 people in a 5% sample from that city. The estimate of the number receiving less than the $2000 wage and salary income would, with a very high probability, when estimated from the 5% sample, lie within the range 9300 to 10,700. This kind of accuracy was sufficient to serve many important purposes and, in fact, was as much accuracy as could be justified in the light of the less-than-perfect accuracy of the responses to the question on income. Not only could estimates be prepared from the sample of what would have been shown by a complete census, but in addition, the range of probable difference between the sample estimate and the result of a complete census could be estimated from the sample! Sampling theory also guides in designing samples to achieve a maximum precision of results per unit of cost.

In considering the advantages of the use of sampling, it may appear to some that the main work involved in taking a census is the time it takes in going from door to door and that, once some questions have been asked at a household, the cost of adding questions would be small whether they were added for a 5% sample or for the total population. Such a presumption is far from true. Suppose, for example, that an additional question about whether a person has a chronic illness adds an average of 20 seconds of work for each person counted in the census. With some 200,000,000 people in the population, this would add more than 1,000,000 hours of work and perhaps $4,000,000 to the cost of the census. Thus, obtaining the added information for, say, five questions from only a 5% sample instead of from all persons can produce needed and highly useful results at a fraction of the cost for complete coverage. Finally, the use of a sample permits tabulation and analysis much sooner than complete coverage.

Starting with the forties, sample surveys on a wide range of subjects were introduced so that continuing and up-to-date information would be available between censuses. For example, the Current Population Survey is a sample of the population conducted monthly by the Bureau of the Census. The Survey collects information each month on employment, unemployment, and other labor force characteristics and activities of the population. It also serves to collect information on other subjects, with different supplemental questions in various months. In one month each year almost the full range of population census questions is asked. In other months information may be requested on recreational activity, housing, disability, or other subjects. Sample surveys include a continuing health survey, and surveys covering retail trade, business and personal services, the activities of governmental units, and so on. These surveys have large enough samples to provide national information, some information for regions of the nation, and even information for large states and metropolitan areas. They cannot, however, provide information for the many individual cities and counties, and for relatively small communities within the cities and counties. To obtain that kind of fine detail, very large samples are needed, samples such as those taken as part of the decennial census.

In the 1950 census, the use of sampling was extended to some questions that in earlier censuses had been collected from all persons. For these questions, a 20% sample was used. This sampling was extended in the 1960 census to most of the items of information. In the 1960 and 1970 population censuses, only the basic listing of the population, with questions on age, sex, race, marital status, and family relationship, was done on a 100% basis.

The following question is often raised: If sampling is so effective a tool, why not take the whole census with a sample? Isn't it a waste of effort to do a complete census? In response, I always point out that the primary purpose of the census is not to obtain national information, but to provide information for individual states, cities and counties, and for small areas within these. The results obtained by converting the whole census to a relatively large sample (perhaps including 20% of the population) *would* be adequate for most purposes. Such a sample, however, would not apportion representatives in the state legislatures in the same way as a complete census. Similarly, there could be important differences in the distribution of vast amounts of funds to thousands of individual small areas. Even very small differences in the total counts by states may decrease by one the number of congressmen from one state and increase by one the number from another—and they can alter the allocation of funds to the states by billions of dollars. Also, in some states, the legal status of many communities depends on the exact size of the official population count; for example, a city with 10,000 or more people can issue bonds. Hence a complete census is needed for total population counts and for some other basic population data, but the great bulk of the information may be collected from a relatively large sample. Much smaller samples can be, and are, used to collect data on items needed only for larger areas such as individual large cities, metropolitan areas, and states.

The Use of Sampling and Experimental Studies to Evaluate and Improve Census Methods and Results. Substantial steps to evaluate and improve census methods were taken beginning in the forties and have been continued and greatly extended since then. Statistical studies have been made of various aspects of the census. One such study was made by repeating the census enumeration, in a well-designed sample of areas, shortly after the original census enumeration, using the same procedures as in the initial census. Thus we were able to see something of how much alike two censuses taken under the same conditions and procedures would be.

Studies of these types show high consistency and accuracy of response for questions such as sex, age, race, and place of birth, but they show higher degrees of inconsistency and inaccuracy in responses to the more difficult questions relating to occupation, unemployment, income, education, and others. The information from such studies guides both in improving the questions in the next census (by, for example, showing which questions cause trouble and need rewording), and in interpreting the accuracy of census results when the questions are put to specific uses.

Studies that compare alternative methods and procedures within the framework of well-designed and randomized experiments have been exceedingly important for learning about the effectiveness of various procedures and in comparing their costs and their accuracies. Such comparisons have been made, for example, of various types of questionnaire designs or of other variations in procedures. For example, one study reexamined the census coverage and questions in a sample of areas, but used more highly trained enumerators, more detailed sets of questions, and other such expensive improvements.

These rather wide-ranging studies led generally to the conclusion that some of the methods earlier regarded as the way to achieve major improvements would not be effective in relation to their cost (although some worthwhile improvements in questionnaire design and procedures were accomplished). We find, for example, that simply spending much more time and money on training an army of temporary enumerators and paying them hourly rates instead of piece rates, making questions more detailed, and insisting on personal interviews for each adult respondent would all add greatly to the cost of a census, but would not make corresponding improvements in its accuracy. There appear to be intrinsic limits to what can be done with a vast temporary organization.

The Surprising Effect of Enumerators. One study, however, led to surprising conclusions and then to a basic improvement in census procedures. It had long been known that enumerators can and do influence the answers they obtain—presumably, unconsciously most of the time. But the *magnitude* of this enumerator effect was not known. Hence a large statistical study was carried out as part of the 1950 census to measure the magnitude of enumerator effect.

The study plan (in a simplified description) was based on areas divided into 16 work assignments (areas small enough so that 2 work assignments could be canvassed by a single enumerator) and 8 enumerators. Two of the 16 work assignments, chosen at *random,* were given to the first enumerator; two of the remaining ones, also chosen at random, were given to the second enumerator, and so on. Of course the whole 16-fold experiment was repeated many times throughout the country.

The random choice of work assignments was essential here in order to interpret the results in a useful way. (Random choice means choice by a method equivalent to writing the numbers 1, 2, . . . , 16 on identical cards, shuffling or mixing them thoroughly, and then picking first one, then another, and so on.)

Another essential feature of the plan was that each enumerator had two work assignments and that there were a number of enumerators. That way, good estimates could be obtained for the variability introduced by a single enumerator (roughly, the differences in performance by the same enumerator that were not attributable to differences in areas) as compared with the variability stemming from differences *among* the enumerators.

Further details of this path-breaking analysis cannot be given here, but we can summarize the results. Far greater differences between enumerators were found than had been anticipated, not so much on items such as age and place of birth, but on the more difficult items such as occupation, employment status, income, and education. For those items, in fact, a complete census would have as much variability in its results (because of enumerator effect) as would a 25% sample if there were no enumerator effect!

What to do about this? One approach might be to expend far more resources on the selection, training, and supervision of enumerators. But the other studies mentioned above had shown that this was not feasible under the conditions of a national census, in which temporary enumerators are hired, trained, and do their job in only two or four weeks.

Another possible answer was to eliminate the need for the enumerator by leaving the carefully designed questionnaires with the respondents and asking them to fill them out and mail them. (Enumerators would be needed only when respondents' returns were incomplete or where the respondents asked for help.) This method was tested in further studies and found to work quite well. It was used in the 1960 census for the larger census forms and was a great success, in terms of both cost and added accuracy. Hence, in 1970, this method was used still further: most of the population received and returned the census forms by mail.

Thus, in the process of using statistics to improve the census, the completeness of coverage has been improved, the accuracy of the items of information collected has been increased, and the time taken to publish the results has been decreased by about half. Now most of the data are available for public use within six months to one and a half years following the completion of the collection of the data, while in the 1950 and earlier censuses the corresponding time for making available the same basic information was roughly one to three years.

With all of this, costs (in terms of equivalent dollars, adjusted for changes in salary rates and other costs) have been reduced. The results have been achieved by the application of sampling and other statistical methods to census data collection in the field, to the processing of the information, and to the study of methods for evaluation and improvement.

PROBLEMS

1. Give several reasons why a census is taken.

2. Why is it desirable in some instances to take a five percent sample as opposed to a complete census? What does the five percent sample lose?

3. Comment on the advantages and disadvantages of using enumerators in a census.

4. Answers to questions regarding sex, age, race and place of birth seem to be more reliable than those regarding occupation, unemployment, income and education. Give two possible reasons for this. Do you think answers to mailed questionnaires would be more accurate than interviews by enumerators on these questions? Why?

5. (For group work). Take a small random sample on some issue, for example, opinions of the local newspaper. Report on the problems and situations that arise.

6. Suppose the Census Bureau decided to add a battery of 6 questions on energy consumption to the 1980 Census. About how much money would the Census save if, instead of asking everyone the questions, they asked them of only a 1% sample?

7. Suppose the 1980 Census uses mailed questionnaires instead of enumerators, and 40% of the returned questionnaires have no answer marked to a question on mental health. Would the Census be justified in counting only the marked responses and publishing the result as a 60% sample? Justify your answer.

8. Someone in the Census Bureau suggests using telephone interviews instead of mailed questionnaires for families with telephones. He thinks that this technique might lead to a more accurate census.
   (a) What arguments can you think of which support or refute his idea?
   (b) Design an experiment to determine whether his idea is correct.
   (c) What other considerations (besides accuracy) would determine whether or not the telephone technique would be adopted?

REFERENCES

John Cummings. 1918. "Statistical Work of the Federal Government of the United States." John Koren, ed., *The History of Statistics.* New York: Macmillan. Pp. 678–679.

Morris H. Hansen, Leon Pritzker, and Joseph Steinberg. 1959. "The Evaluation and Research Program of the 1960 Censuses." *1959 Proceedings of the Social Statistics Section, American Statistical Association*. Washington: American Statistical Association.

Morris H. Hansen and Benjamin J. Tepping. 1969. "Progress and Problems in Survey Methods and Theory Illustrated by the Work of the United States Bureau of the Census." Opening address at Symposium on Foundations of Survey Sampling, April 22–26, 1968, University of North Carolina, Chapel Hill. Norman L. Johnson and Harry Smith, Jr., eds., *New Developments in Survey Sampling*. New York: Wiley (Interscience). Pp. 3E(1)–3E(11).

Ann Herbert Scott. 1968. *Census, U. S. A.: Fact Finding for the American People 1790–1970*. New York: Seabury.

Joseph Waksberg and Robert B. Pearl. 1965. "New Methodological Research on Labor Force Measurements." *1965 Proceedings of the Social Statistics Section, American Statistical Association*. Philadelphia: American Statistical Association. Pp. 227–237.

HOW CROWDED WILL WE BECOME?

Nathan Keyfitz *University of California, Berkeley*

All statistical facts concern the past. The Census of April 1970 counted 205 million of us, but we did not know this until November, despite the census emphasis on speed, pursued with ingenuity and with much new electronic equipment. Stock-market prices and volumes are hours old before they appear in the evening paper. Statistics of plans or intentions are only an apparent exception. No one can ever gather data directly on the future.

Yet the actions that statistics serve to guide can occur only in the future. The local telephone company wants to know how much this town will grow in population over the next few decades. Its interest is not abstract curiosity, but contemplated construction of new lines out toward a certain suburb. The investment might occur in the next two or three years, and the service given by the investment along with the income derived from it would be spread over 30 years. If the town does not grow as much as expected, the construction would be wasteful. If the growth is in the direction of a different suburb, then lines will be idle on one side of the town and too often busy on the other side. School authorities, the bus company, a textile manufacturer, all similarly need statistics on the future for the conduct of their business, and these are nowhere to be collected until the future has become past and it is too late.

With producers of population statistics all working on the near side of *now* and users all concerned with the far side, it is lucky that even in times of rapid change, some continuities are to be found between past and future. Population projection rests on these continuities.

The continuities are not be found in simple totals. We know that the number of people in the U.S. does not increase evenly from year to year, and still less does the population of one town or one age group increase evenly. The age classes especially have fluctuated erratically in recent decades. Today the U.S. includes an exceptionally large proportion of young people 10 to 25 years of age, the result of the baby boom of the forties and fifties. They have crowded the high schools and colleges, and they are seeking jobs and entry into graduate schools across the country. But during the sixties, births fell sharply, and the number of pupils entering elementary schools leveled off.

Yet we can say something about the future. At the end of the seventies, schools and the labor market will be reached by the wave of what may be called the nonbirths of the sixties. But, though kindergartens and public schools will slow their expansion in the seventies, they may have to accelerate it again in the eighties to accommodate a new generation—children of the children born in the postwar baby boom. How such things can be projected with some confidence is our subject.

The approach, or model, that we shall build for projection serves other purposes than prediction. It is especially valuable for judging the effects on population growth of a possible change or a proposed policy.

PROJECTION WITH CONSTANT BIRTH AND DEATH RATES

The trick in projection is to seek elements that remain nearly constant through time. The increase in total population from year to year plainly does not qualify, but certain *rates* do remain more or less the same, and on these we rest our analysis of the future. For example, the proportion of people aged 30 who die each year is likely to remain much the same in 1960, 1970, and 1980. These death rates are constant enough that some fairly reliable predictions can be hung on them, and we proceed to the exploitation of this constancy.

Our projection of population into the future includes three parts:

(1) The statistical data of a baseline census from which work starts
(2) Effect of death
(3) Effect of birth

Demographers ordinarily recognize five-year age groups, to the end of life, for men and women separately, and they have a computer do the arithmetic. To show the procedure without being swamped in numbers, we consider here girls and women only, and these just up to age 45. Moreover, we need only consider three age groups, each of 15 years' width. For purposes of this illustration, three numbers describe the population at any one time.

We can make a fairly complete analysis for these three groups, and show the whole worksheet. The census of April 1, 1960, counted 27.4 million girls under 15 in the U.S. It showed only 17.7 million between 15 and 29 years. An intermediate number, 18.4 million, were between 30 and 44. (This article follows the census in always counting people at their age last birthday.) Those under 15, born between 1945 and 1960, constitute the baby boom; the next older group, born between 1930 and 1945, are survivors of the meager crop of depression babies; the oldest, aged 30 to 44, were born between 1915 and 1930, when birth rates in the U.S. were higher than in the thirties, but lower than in the fifties.

Now these three numbers can be written one below another in an array known as an age distribution; see Table 1.

So much for the counts made in 1960, our point of takeoff into the future. We now need to know how death and birth will act on this starting distribution. (Migration, which demographers usually take into account in making projections, is probably going to be relatively small and not likely to affect our conclusions seriously, so we shall ignore it.)

Let us start with death, but look at its positive side: the people who do not die, but survive into the next period. The question is, how many of the 27.4 million girls under 15 years of age counted in the 1960 census may be expected to survive to 1975? We have at hand a *life table,* as such collections of survival probabilities are called, that gives the proportion of girls under 15 who survive for 15 years as approximately 0.9924. This life table was calculated from deaths in the U.S. in 1965, and it would not be very different if calculated for any other recent year. Hence the expected number of survivors 15 years later of the 27.4 million counted in 1960 would be

TABLE 1. Age Distribution of American Girls and Women, 1960

AGE	MILLIONS OF GIRLS AND WOMEN
0–14	27.4
15–29	17.7
30–44	18.4

27.4 multiplied by 0.9924, or 27.2 million. These girls would be 15 to 29 years of age in 1975.

No such multiplication can give the *exact* numbers in 1975. Individuals survive or die at random, and even if 0.9924 were the probability for each separate girl 0–14 years of age, a few more or a few less than 27.4 × 0.9924 million could survive in the particular years 1960–75. If the U.S. were subject to serious epidemics, chance events each affecting large numbers of people, then the variation from year to year would be substantial. Because, in fact, death and survivorship act like events affecting each of us more or less independently, the multiplication is permissible, though even then the result could be made wrong by a war or epidemic on the one hand or a medical breakthrough on the other. We shall suppose that the chance of survival does not change very greatly over the period of the projection.

In the same way the proportion surviving 15 years among girls 15 to 29 in 1960 is estimated at 0.9826, and hence the projected number aged 30 to 44 in 1975 would be 17.7 × 0.9826 = 17.4 million. The projections to this point stand as shown in Table 2. Our next task is to fill the upper cell on the right, which requires an estimate of the number under 15 in 1975. (Remember that, to keep things manageable and simple, we are neglecting women 45 or more—of course, only for present simplicity, as the wives of some of us will remind us.)

All of the girls under 15 years of age in 1975 will have been born since 1960, and we need to estimate not how many girl births take place in the 15 years, but how many of these births survive to 1975. We know, also from the 1965 experience, that, on the average a woman 15 to 29 can expect 0.8498 surviving girl babies by the end of a 15-year period. We have counted girl babies only for this purpose because a female model is what we are constructing, and we have deducted deaths among the babies so as to come up with girls under 15 who will be alive in 1975. There were 17.7 million women aged 15 to 29 in 1960, and their contribution to the total girls under 15 in 1975 is expected to be 17.7 × 0.8498 = 15.0 million.

TABLE 2. Projected 1975 Population of American Girls and Women

AGE	MILLIONS OF GIRLS AND WOMEN	
	1960	1975
0–14	27.4	?
15–29	17.7	27.2
30–44	18.4	17.4

Children will be born also to the women 30 to 44 years of age; on the average, these women will have 0.1273 girl babies alive at the end of the 15-year period. The contribution that these make to the total girls under 15 in 1975 is expected to be 18.4 × 0.1273 = 2.4 million. (The actual calculation was made to more decimals than shown here.)

Finally, children will be born before 1975 to girls under 15 in 1960, a large proportion of whom will become of childbearing age during the 15 years. On the average (again at 1965 rates), they will have 0.4271 surviving girls. This average, like the others above, is taken over many different cases; it includes the girls too young to become mothers, those who will be old enough but not yet married, and those who will marry but not have children. The expected contribution here is 27.4 × 0.4271 = 11.7 million.

To find the total number of girl children under 15 surviving in 1975 we must add the numbers reached in the three preceding paragraphs: 11.7 + 15.0 + 2.4 = 29.1 million in all. Figure 1 shows schematically what is happening. (Because so few children are born to women over 44, we can afford to ignore them. Our simple model will give almost the same rate of increase of the population as more elaborate models.)

By repeating exactly the same argument, except that we now start with the 1975 projected population, we obtain the age distribution in 1990; any

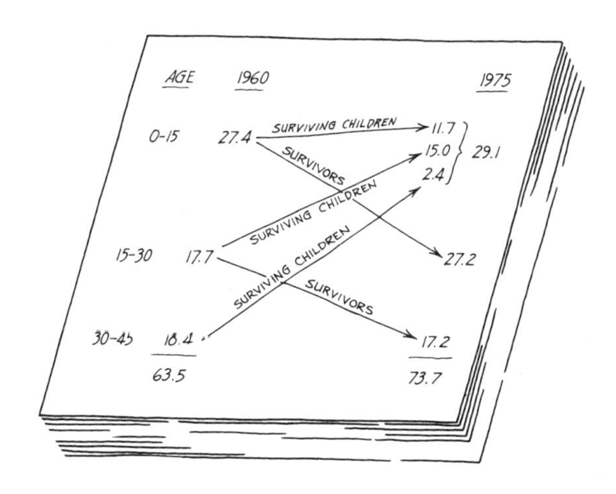

FIGURE 1

Calculation of 1975 population of girls and women under 45 years of age (figures are in millions)

TABLE 3. Millions of Girls and Women Under 45 Years of Age in the U.S. if Birth and Death Rates Remain at the 1965 Level

AGE	1960	1975	1990	2005	2020	2035	2050	2065
0–14	27.4	29.1	37.7	44.1	54.3	65.0	79.0	95.3
15–29	17.7	27.2	28.9	37.5	43.7	53.8	64.5	78.4
30–44	18.4	17.4	26.7	28.4	36.8	43.0	52.9	63.4
Total	63.5	73.7	93.3	110.0	134.8	161.8	196.4	237.1

number of additional 15-year cycles may be calculated similarly. Table 3 shows the resulting numbers up to 2065.

WAVES OF MOTHERHOOD

The first age group, girls under 15, increases less than two million between 1960 and 1975, while the women 15 to 29 increase by almost 10 million. The 15 to 29 group in 1975 are the babies born between 1945 and 1960, the postwar baby boom, and as these succeed the depression babies in any group we expect its number to rise rapidly. Women 30 to 44 actually become fewer during this first 15-year period, even though the 0 to 44 population as a whole is growing.

Because most children are born to mothers 15 to 29 years of age, we can expect a new baby boom, an echo of the first one, at the time when the babies of the fifties themselves pass through childbearing age, and indeed the under-15s grow by 8.6 million from 1975 to 1990 according to Table 3.

In fact, the depression and boom will keep echoing to much later times, supposing, as we do throughout, that childbearing practices remain fixed. But the table also shows that as time goes on the irregularity of the 1960 age distribution steadily lessens. At the end of 105 years all ages are increasing at very nearly the same rate.

That the several ages ultimately increase at the same rate can be seen by dividing each 2065 figure in Table 3 by the corresponding 2050 figure. In Table 4, this ratio is shown to be about 1.2 for the three age groups and the total. By carrying the projection further, we could have had these ratios as close to one another as we wanted; in fact, further calculation shows that they all would converge to 1.2093.

This ratio may be called intrinsic, or the true ratio of natural increase. It can be shown to depend not at all on the 1960 age distribution with which the process started, but only on the rates of birth and death, and it is the most informative single summary measure of that set of rates. It tells us that any population that is subject to our particular birth and death rates

TABLE 4. Increase of Age Groups of Girls and Women in the U.S. from 2050 to 2065

AGE	2050 (MILLIONS)	2065 (MILLIONS)	RATIO, 2065 TO 2050
0–14	79.0	95.3	1.206
15–19	64.5	78.4	1.216
30–44	52.9	63.4	1.198
Total	196.4	237.1	1.207

over a period of time will sooner or later settle down to an increase in the ratio 1.2093, which is to say by about 21% per 15-year period. Under the operation of the projection, applying the assumptions we have made, *a stable age distribution* is sooner or later attained in which all the irregularities of 1960 due to boom and depression have been forgotten. Age distributions tend to forget their past when persistently pushed forward by the method developed above.

Let us find numerically the component of population growth that increases in the same ratio in every cycle, a mode of increase spoken of as *geometric*. If we divide each of the numbers shown under the year 2065 in Table 3 by 1.2093, we get back to an estimate for 2050; if we then divide again by 1.2093 we get back to 2035, and so on. To get back to 1960 we would divide by the seventh power of 1.2093, written $(1.2093)^7$ and equal to 3.78. Carrying out the division gives 95.3/3.78 or 25.2 million for age 0 to 14, and similar calculations for the other ages provide what we may call the stable equivalent for 1960; see Table 5.

Table 5 shows the set of numbers that, increasing in the constant ratio 1.2093, would sooner or later exactly join the track of our projection in each age group. If we multiply the stable equivalent by the fixed number 1.2093 to obtain the geometric track, and subtract this from the projection of Table

TABLE 5. Main Component of Female Population in the U.S., 1960

AGE	STABLE EQUIVALENT (MILLIONS)
0–14	25.2
15–29	20.7
30–44	16.8
Total	62.7

TABLE 6. Departures of Projected Population in Table 3 from Geometric Progression in Millions

AGE	1960	1975	1990	2005
0–14	2.2	−1.4	0.9	−0.6
15–29	−3.0	2.2	−1.4	0.9
30–44	1.6	−2.9	2.1	−1.4

3, we obtain Table 6. For example, for girls 0 to 14 in 1975, we have $29.1 - 25.2 \times 1.2093 = -1.4$. Our analysis has separated the prospective population change into two parts, one a smooth geometric increase, the other a series of waves that are departures from the geometric.

These departures gradually diminish in amplitude. For 1960, we have 2.2 million as a measure of the temporary "excess" of the 1945–60 babies. The −3.0 million are the deficiency of the depression babies, and 1.6 million, again an excess, relate to the twenties. Note that by 1990, each of these has an echo, of the same sign but on the whole of smaller amount.

The tendency of the waves to diminish in amplitude is related to women having their children over a range of ages. If all children were born to mothers of the same age, the waves would steadily *increase* in amplitude. With such concentration any irregularity in the age distribution caused, for example, by a war or depression would not only continue echoing through all later generations, but become magnified. In the U.S. today, women prefer to have their children around age 25, whereas our grandmothers spread theirs from about 20 to 45. The new style, associated with the effective use of birth control, could mean diminished stability.

In this analysis of the U.S. population we have gone from the facts of the 1960 census, through various more or less realistic calculations concerning 1975 and even 1990, into a kind of fantasy as we proceed far into the future. The early part of the projection can within limits be useful for practical purposes; the later part is so dependent on various *if's* that one would be very foolish to count on it. The biggest doubt attaches to the birth rate. It may seem that birth is as individual a matter as death, and therefore births across the country ought to be independent of one another, yet in fact high and low birth rates spread like epidemics across the country.

Why then do we bother with the fantasy of such far-out projections referring to the distant future? The answer is that they help us understand the present. We ascertain the meaning of the present rates of birth and death by calculating what they *would* lead to if they continued for a hundred years or more. Let us see why this is even more important in study of the birth and death rates of developing countries than of a developed one like the U.S.

GROWTH OF DEVELOPING COUNTRIES

The task is in some ways easier for developing countries because they do not have a history of changing birth rates. It is true that their death rates have been falling, and where this occurs for young children only, it is the equivalent of a rise in the birth rate: as far as the population mathematics is concerned, a fall in infant deaths has the same effect as a rise in births. In fact, however, deaths have been falling at nearly all ages, and births are relatively unchanged. The fact is that age distributions are already more or less in the condition we called stable, and which could be attained by the U.S. only in the course of several generations of fixed rates. Because of past uniformly high birth rates, developing countries tend to grow much faster than the U.S. Moreover, they show a simple geometric increase, with all ages rising uniformly. The sort of waves that we have been studying do not occur for them.

Let us concentrate then on the geometric component, and take Malaysia as an example. In the mid-sixties, Malaysia was growing in the intrinsic ratio defined above of 1.59 per 15 years. This corresponds to an annual rate of increase of the 15th root of 1.59 or 1.031, that is, about 3.1% per year, against about 1% for the U.S. To convince ourselves of this we could multiply 1.031 by itself 15 times, that is calculate $(1.031)^{15}$, and we would find the result to be just under 1.59.

We can see the long-term prospect more clearly by translating into doubling times. How long does it take a country that is increasing at the rate r % per year to double in population? The equation to be solved for the unknown time t is $[1 + (r/100)]^t = 2$. The solution is obtained by taking logarithms of both sides and comes out very near to $t = 70/r$, where r is expressed as percent increase. This rule applies to money lent out at interest, and financiers use it because they are very interested in doubling times. The same sort of rule works for the half-life of a piece of radium or other substance under radioactive decay.

As an example of a geometric projection of a population, suppose that Malaysia's rate of 3.1% per year were to go on for about $70/3.1 = 23$ years. This would carry it from the present 10 up to 20 million people. Suppose it went on for another 23 years; this would mean another doubling. At the end of 115 years at this rate, the population would have doubled five times, which means multiplying by 2^5 or 32; Malaysia would contain 320 million persons. By the end of 230 years, it would have doubled ten times and would contain 2^{10} times as many as now, or 1024 times ten million.

No one could mistake such calculations for predictions of what will happen. In a sense they are the opposite; we might call them counterpredictions, for they show that in much less than 100 years, the birth rate will go down or the death rate will go up or both. Most demographers are optimistic enough to believe that the adjustment will be through the birth rate.

Other countries are today growing faster than Malaysia. Mexico's present 50 million population is increasing at about 3.5% per year, which, by our rule, would give it a doubling time of $70/3.5 = 20$ years. At this rate, it would be 100 million by the year 1990, 200 million by the year 2010, and 400 million by the year 2030. This again is a counterprediction; shortage of food, excess of pollution, and many other reasons would prevent it from coming true. The usefulness of the calculation is in showing that births must be reduced; anyone who makes a *principle* of permanent opposition to birth control in effect favors an increase of the death rate sooner or later. The most that opponents of birth control can argue is that it should be delayed a few years.

In diluted form, the same is true of the U.S. Our calculation showed that the geometric component, neglecting waves, was an increase of 21% for 15 years, or about 1.2% per year, according to births and deaths of 1965. That means a doubling in 60 years, quadrupling in 120 years, and so on. Contract the U.S. time scale by about three, and the future growth of the U.S. is the same as that of Mexico. And even our having three years to Mexico's one is partly offset by the greater damage to the environment caused by our more advanced industry.

RELIABILITY OF PREDICTION

The techniques presented in this article and obvious extensions of them are much used for predicting the future. They are used not because they are perfect, but because nothing better is available. Whatever continuities exist in birth and death rates are exploited by the makers of projections. From about 1870 to 1935 in Western Europe and the U.S., the birth rate and the death rate were both falling; projections could be made by the method outlined here, except that instead of using fixed rates, the past downward trend in birth and death was projected into the future. Such projections were acceptably accurate as long as the downward trend continued.

But these same countries reached a turning point in the forties. People married younger, and births rose rapidly. Moreover, couples varied the timing of their children as well as varying the total number. The fact that in a modern society couples plan their children, both in number and in timing, can be used to strengthen the predictions. Samples of young couples are surveyed to find what their childbearing intentions are, just as we ask intentions on buying houses and automobiles. The official estimates of the U.S. Bureau of the Census take account of these intentions.

The Census Bureau's projections, which use a vastly elaborated form of the method of this article, can be compared with ours. Theirs are more detailed than ours above have been, and they are also cautious enough to make a variety of projections rather than betting on just one. They end up with four numbers for each age, sex, and future year. For example, for 1990, their four numbers for girls 0 to 14 years of age range from a low of 29.5 million to a high of 41.8 million. Our Table 3 shows 37.7 million.

How well would past application of our model have foretold the 1970 population of the U.S.? If we had worked forward from the 1920 census total of 106 million, using exactly the procedure of this article, but applying it in five-year age intervals to all ages and to both sexes, we would have found about 185 million for 1970. If we had allowed for immigration less emigration of 200,000 per year, this would have brought us to 195 million against the 205 million actually counted. Something a little lower would have been found starting from 1950; starting from 1960, we would have slightly overestimated the census figure. An error of about 5% in estimates made up to 50 years ago is not bad, considering that the Bureau of the Census estimates its own actual count to be subject to nearly 2% error.

We would have done much worse starting in 1940, however; the method of this essay, plus about 6 million immigrants, would have produced a total of only about 160 million. Put another way that sounds even worse: the increase from 1940 to 1970 was about 62 million, and of it, we would have estimated less than 30 million. This is not a good score. The baby boom of the fifties was a historic event about as hard to predict in advance as the war that sparked it.

MODELS PERMIT EXPERIMENTS

We have discussed the population projection as a way of making predictions, and also as a way of making counterpredictions—calculating what would happen if present rates continued as a way of showing that they cannot continue. This last suggests what may be the most important use of the model of this essay, originally developed and still often applied to making predictions. This use is experimentation. Not only does our model answer the question "What would happen if the birth and death rates of the present time continue into the future?" but it answers a great variety of other important questions. What would be the effect on total population numbers if intensive research on heart disease was undertaken, and it reduced deaths from that cause by 90%? This could conceivably result from a research effort comparable to present investigations of outer space. But the effort could equally be put into reduction of infant mortality. Our model could compare the effects of these alternatives, taking account of the fact that the person dying of heart disease is of such an age that he will soon die of something else; the child saved from some lethal ailment, on the average, will have a long life ahead of him. A given fall in infant mortality increases the population by much more than the same fall in heart disease.

An example of a question that has been frequently asked, and to which our model provides a clear answer, is: what would happen if, starting now, each member of the population averaged one descendant? This means that each fertile couple would need to have somewhat more than two children, to allow for those who do not marry, for those who marry but are infertile, and for deaths in childhood. An average of about 2.3 children per married couple would constitute bare replacement, that is to say, would keep the population stationary at modern death rates.

But the stationary *level* at which it would keep it would be above that of the starting point. Any population that has been subject to birth rates higher than bare replacement in the past has a large proportion of girls and women of childbearing age. These will produce increasing numbers of children for about 50 years after the date at which the birth rate falls. The projection model developed in this article tells how high the population would rise if we drop to bare replacement numbers of children.

Application of the model shows that the U.S. would rise to about 270 million persons if bare replacement were adopted today, and Mexico would rise from its present 50 million to over 80 million. Most underdeveloped countries, such as Mexico, would increase by about 65% from the point at which they drop to replacement, and they would do this over 50 or so years. No country ought to fear that immediate adoption of contraception would freeze total population where it stands; a kind of momentum operates simply because of the favorable age distribution that results from past high fertility.

Former President Sukarno of Indonesia was against birth control because he thought Indonesia should have 250 million people. The present government applied the model described in this essay and found it will probably exceed 250 million even if the brakes are put on immediately; consequently it has formally sponsored a program of birth control.

CONCLUSION

We started this essay by developing a model to forecast the future. The model works for forecasts over short periods and over longer periods in which either the trend is steady or in which ups and downs offset one another. For forecasting major turning points it is of little use, but so is any other model so far developed.

While the model is moderately, but only moderately, successful for the purpose for which it is designed, it has the power to analyse hypothetical futures whose consideration is urgent. If the marriage age in India is raised to 20, what effect will this have on the birth rate? If 20% of couples aged

30 accept sterilization, how far will this take a given country towards zero population growth? How much long-run effect does the emigration from Jamaica have on its population increase? What kind of population waves would follow a sudden drop in the birth rate of an underdeveloped country? It is in answering questions such as these, of which examples have been given in the course of this essay quite as much as in making predictions, that the projection model presented here finds its use.

PROBLEMS

1. Briefly explain the meaning of the phrase, "geometric population increase."

2. Suppose that country A has been growing at a geometric rate for a number of years. If the rate were suddenly to drop to the bare replacement rate, the population would continue to increase for a while. Explain why this is so. (Remember that the bare replacement rate is that at which each member of the population averages one descendant.)

3. Refer to the "Growth of Developing Countries" section of the article. How long will it take a country that is growing at a constant rate of 1.5% a year to double in population?

4. Suppose a country's population is increasing at a 2% annual rate. If this rate were maintained, and the 1975 population were 50 million, how large would the population be in 2045?

5. An animal breeder has developed a new strain of rodent, the females of which produce a beautiful pelt in 3 years. The breeder finds that about 80% of the rodents survive into their second year, and 60% of the rodents in their second year survive into their third year. He also finds that on the average, females in their first year produce .5 female offspring; females in their second year produce 2 female offspring; and females in their third year produce .5 female offspring.

In 1975, the breeder has 400 females in their first year, 200 in their second year, and 100 in their third year.

Assuming that the death and birth rates remain constant, how many 3 year-old pelts would you predict the breeder can harvest in 1978?

6. Why was the Bureau of the Census population estimate for 1970 based on the 1920 Census better than the estimate based on the 1940 Census?

7. Look at Table 6.

(a) Why does every other entry on each row have a minus sign?

(b) Why do the minus signs in the second and third rows appear in different columns?

THE CONSUMER PRICE INDEX

Philip J. McCarthy *New York State School of Industrial and Labor Relations*
Cornell University

ALMOST EVERYONE worries about the prices he pays for all kinds of things. Perhaps the best way for an individual to find out how the price level changes is to follow in newspapers the Consumer Price Index (CPI) of the Bureau of Labor Statistics, a part of the U.S. Department of Labor. The individual consumer sees the CPI as a good measure of price changes in goods and services that he purchases. He reacts to newspaper statements such as "In September of 1969, the average urban family must spend $12.93 for the same amount of goods and services that could be obtained for $10.00 in 1957–59." He wonders whether or not his income has increased sufficiently to compensate for this increase in prices.

Unions and management pay particular attention to the CPI because they know its value will play a critical role in wage agreements and that a 7% annual increase, for example, in the CPI will lead to a demand for at least

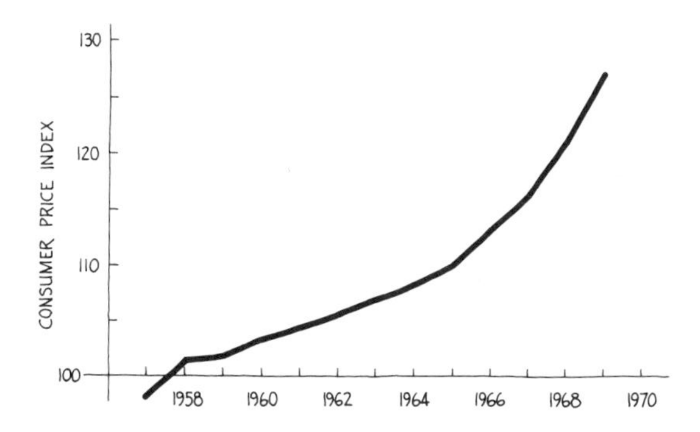

FIGURE 1
Consumer Price Index, annual averages (1957–59 = 100)

a 7% increase in wages. Furthermore, in January of 1969 there were 2.55 million workers whose wages were covered by contracts containing *escalator* provisions, that is, provisions calling for automatic changes in wage rates in accordance with specified changes in the CPI.

Finally, economists concerned with the fiscal and monetary policies of the U.S. Government use the CPI as one of the principal indicators of the existence of an inflationary spiral in which higher prices lead to higher wages, which, in turn, lead to a greater demand for goods, thus raising prices again, and so on. During recent years, most writings on the state of the U.S. economy have emphasized the fear of inflation as evidenced by the increasing values of the CPI. The reason for this fear is illustrated clearly in Figure 1, which shows annual average values of the CPI for 1957–69. These values have been rising at an increasing rate since about 1964.

Although statistical studies of prices and living conditions in the U.S. were conducted in the late nineteenth and early twentieth centuries, the first complete "cost-of-living" indexes were published by the Bureau of Labor Statistics in 1919. They referred to 32 large shipbuilding and industrial centers, and arose through an agreement between the Shipbuilding Labor Adjustment Board and labor chiefs that one of the factors to be considered in settling labor disputes was that of "adjusting wages to the higher cost of living resulting from the war."

Since that time, the CPI has broadened in scope and increased in importance. This has led to many professional appraisals of the CPI, among them one by an advisory committee of the American Statistical Association in 1933–34 and one by the National Bureau of Economic Research in 1959–60. These appraisals have influenced major revisions in the CPI, made at approximately ten-year intervals since 1940, as well as many minor revisions.

PROBLEMS IN CONSTRUCTING THE CPI

The typical consumer is quite aware of changes that occur in the prices of goods and services that he purchases regularly. He knows when the price of gasoline increases by one cent per gallon, or when the price of milk is increased by one cent per quart. Furthermore, he is able to predict with reasonable accuracy the impact of these price changes on his monthly or yearly budget, and he is therefore in a position to estimate the extra income (including an allowance for taxes) he must obtain in order to compensate for these increases. Price changes relating to less frequent and more sporadic purchases—clothing, doctors' services, appliances, automobiles, and homes, for example—are not as visible to the individual consumer, and it is much more difficult for him to assess their effects on his budget. Thus he may well know that it costs him more to live, or less to live, during the current year than in the preceding year, but he would find it almost impossible to provide an exact value for this change in his cost of living. His problem would be further complicated by changes that might occur in such matters as family status (births and deaths), family desires (color TV instead of black and white), and family purchasing patterns (chicken instead of steak). And yet it is exactly this evaluation—for the nation as a whole, for separate regions and cities of the nation, for different classes of expenditures (food, clothing, housing, transportation, and the like), and on a monthly basis—that the CPI provides. This essay is concerned primarily with the contribution of statistics to the construction of the CPI.

If the individual consumer actually wished to keep sufficiently detailed records to measure the change in prices of the items he purchases from, say, January of 1968 to January of 1969, he might proceed as follows. During January of 1968, he would keep a record of every purchase and would then summarize these purchases in terms of quantities (numbers of quarts of milk, number of pairs of shoes, number of haircuts, and the like) and the unit price of each item. In effect, he defines a "market basket" of goods and services and obtains the cost of this market basket in January 1968. Suppose this cost is $750.00. In January 1969, he would price exactly the same market basket of goods and services. Let us assume that the cost then was $800.00. Therefore, this consumer's personal CPI for January 1969 with January 1968 weights (quantities) and with January 1968 as base (= 100) is $100 \times 800/750 = 107$. The CPI is effectively computed in the same way, but the operation is much more complicated, since it deals with a large population rather than a single consumer.

Even if we assume that the family status and desires of our consumer did not change from January 1968 to January 1969—an assumption that can seldom be exactly true—there are many practical problems and tantalizing questions that would plague those responsible for the construction and interpretation of this index. In particular,

(1) Even for a single consuming family, the number of different items purchased during a month or year is extremely large, and the continual pricing of such a list would be costly and time-consuming. Preparation of a CPI for groups of families obviously magnifies this problem to a wholly impractical size.

(2) The average family has available to it a host of vendors of goods and services, and prices vary from vendor to vendor. In January of 1969 should one attempt to return to the same places where purchases were made in 1968? And what about stores that have gone out of business or stores that have come into existence in the interim?

(3) Prices vary from day to day. Must our consuming family return to the same store at the same time to compute the cost of its market basket in January of 1969? And how does one take into account the facts that some families take advantage of sales and others do not and that some families shop on a day-to-day basis while others shop less frequently?

(4) Suppose that a major purchase, such as an automobile, was made by our family in January 1968. The fixed market basket still contained this item in January 1969 even though such a purchase was not made at that time. How should this be handled?

(5) The goods and services available to the consumer change from time to time. Items contained in the original market basket may not exist in the stores when we return to price them at a later date, or they may exist only at an improved or lowered quality level. What do price changes mean under these circumstances?

(6) We also observe that our consuming family—even though its composition, status, and desires are assumed not to have changed—may be able to "beat" a rise in prices by appropriately altering its purchasing patterns. Suppose, for example, that they notice that the price of steak has risen sharply and decide that the desires of the family are as well satisfied by chicken as by steak. If chicken is substituted for steak, the family's satisfaction level will not change, and yet the amount spent for food may actually decrease in spite of what may be a general increase in the price level. In effect, we can think of replacing the "fixed-contents" market basket with a "fixed-satisfaction" market basket.

The above problems relate to both measurement and concept. Thus we may find that a particular item of merchandise has changed in quality, and yet it may be a most difficult task to measure this change in quality and to translate it into a dollar value. The Bureau of Labor Statistics does attempt to account for these changes in quality in constructing the CPI.

Item 6 above is at a deeper conceptual level. Most economists would prefer a "constant-utility" or "constant-satisfaction" or "welfare" type of price index to the "fixed-market-basket" type of index that is currently produced by the Bureau of Labor Statistics. In other words, what change in expenditures must a consuming family make from one time to another in order to maintain a constant level of satisfaction, with the recognition that the contents of the market basket can be changed in order to accommodate changes in prices, changes in products and the like? Although steps are being taken to move the CPI in the direction of such a "constant-utility" index, the practical and conceptual problems are difficult to overcome, and progress has been slow. As a matter of fact, in September 1945, the official title of the index was changed from "The Cost of Living Index" to "The Consumer Price Index" in order to emphasize the distinction between these two approaches. In this essay we shall not treat these conceptual problems, but rather we shall focus on the role of statistics as it helps to solve CPI measurement problems.

SAMPLING AND THE CPI

Many of the above problems become more manageable if attention is shifted from the individual consuming family to a group, or *population,* of consuming families. Thus, even though some members of the population did purchase automobiles in January 1968, other members of the population did not make such a purchase, and similarly for January 1969. An automobile can then be introduced into the market basket with a weight that will reflect the effects of changes in its average price, averaged over the population of families. This shift in emphasis from the individual consumer to a population of consumers, where differences among the members of this population are known to exist, means that the construction of the CPI requires statistical sampling (and analysis) from the population of consuming families.

Moving in this direction also forces one to think in terms of the *population of goods and services* available to all members of the population of consumers and in terms of the *population of outlets* at which all of these goods and services may be purchased. Furthermore, it is manifestly impossible to study every consuming family, and to price each item of consumption in all the outlets where it can be purchased. Hence it becomes necessary to select samples from each of these populations and to draw inferences from the samples to the entire populations. The methods of statistics can assist and guide these steps.

The Consumer Price Index has never attempted to measure the changes in the prices of goods and services for all families and individuals living in the United States. Rather, because of its traditional use in collective bargaining between labor and management, its scope has been restricted to urban families. More specifically, the *population of consuming families* covered by the index consists of all urban families (including single workers) for which 50% or more of the family's income comes from wages or from salaries earned in clerical occupations and for which at least one member of the family unit works for at least 37 weeks during a year.

In determining the contents of an average, or "representative," market basket of goods and services for this defined CPI population of consuming families, it is impossible to study every family. Hence a sample, or a portion of this population, must be chosen. Because a list of the members of the population is not available, complex methods of statistical sampling must be employed.

In brief, a sample of urban communities is selected first, and then a sample of families is taken from each of the selected communities. In the 1959–64 revision, this was accomplished through an original selection of 50 cities. The 12 largest cities in the U.S., plus one city from Alaska and one city from Hawaii, were automatically included in this sample of cities. The remaining urban areas of 2500 and over were placed in homogeneous groups according to size and geographic location and 36 cities were chosen from these groups in accordance with probability methods of selection. An additional six large cities were added to the sample later, primarily so that individual city indexes could be published. This sample of 56 cities serves not only as a basis for studying the expenditure patterns of consuming families, but also as a basis from which to obtain the prices that must be used to determine the current cost of the CPI's market basket of goods and services. The CPI will continue to be based upon this sample of 56 cities until the next major revision takes place.

Within each of the 56 chosen cities, a sample of consuming units was selected and interviewed during the period 1960–61. These units were drawn in accordance with the tenets of statistical sampling theory; the goal was to choose samples in such a way that objective measures could be devised for assessing the likely size of deviations between averages for the sample and corresponding (unknown) averages for the whole population. This particular survey was called the *Consumer Expenditure Survey* (CES) because its primary purpose was to collect data relating to family expenditures for goods and services used in day-to-day living. Information was obtained through lengthy personal interviews with family members. Among the items recorded during this interview were the following:

(1) A complete account of receipts and disbursements for the preceding calendar year.

(2) The estimated value of goods and services received free.

(3) The characteristics of housing occupied by both home owners and renters.

(4) An inventory of major household furnishings owned.

(5) A detailed listing for a seven-day period of expenditures for food and beverages, household supplies, and tobacco.

Altogether, intensive interviews were conducted with 9476 consuming units. Of these, 4860 interviews were with members of the defined population of consumers for the CPI. The remaining 4616 units did not satisfy the CPI restriction to the population of consuming families, although the data obtained from them are of value for other purposes.

Data collected in the CES interviews with members of the CPI population were used for a variety of purposes in determining the structure of the current CPI. First of all, these data determined the complete contents and total cost of each family's market basket in the survey year. The number of items of expenditure in all market baskets totaled about 1800. The items were classified into five major groups: food, housing, apparel and upkeep, transportation, and health and recreation. Each of these groups was further subdivided, so that the final classification scheme consisted of 52 expenditure classes. Some examples of expenditure classes are: meats, eggs, fuel and utilities, housekeeping services, footwear, auto repairs and maintenance, hospital services and health insurance, and reading and education. In effect, the consumer's market basket was divided into 52 compartments. The total cost of each compartment was then determined for all sample consumers selected from within a particular city, and these total costs were expressed in relative terms; for example, at December 1963 prices, meats were estimated to account for 4.45% of the total cost of the nationwide market basket. Different market baskets are used in different cities to allow for differences in such characteristics as climate and the availability of different foods. Thus we use many average market baskets rather than a market basket that applies to a single consuming unit.

As observed earlier, the CES interviews provided a market basket filled with some 1800 items that consumers had purchased. It would be impossible to price all of these items in their almost infinite variety each month, even recognizing that this pricing need be carried out only in the 56 index cities. Hence another sampling problem arises. Again the theory of statistical sampling was used to select a sample of items from each of the 52 expenditure classes. The final sample for the 1959–64 revision contained 309 items. The sampling approach allows the contents of the market basket to change. For example, if one item disappears from the market, replacements may be made by further sampling. The compartments and their original weights, however, remain fixed through time.

Once a sample of items has been selected and its members specified in detail, another, more difficult sampling problem must be faced. Within each of the 56 index cities, prices of the sample items must be obtained on a monthly basis from the outlets patronized by members of the CPI population of consuming units. It is impractical to obtain price quotations from all possible outlets, and so a sample of outlets must be selected to serve as sources of price information.

Among the problems that have to be considered in developing a sample of outlets are the following:

(1) Ideally, we would define a separate population of outlets for *each* item to be priced and would select a sample from *each* of these outlet populations. This approach would be too costly because even a sample of only four or five outlets for each item would require that 1500 outlets in a city would have to be priced. Furthermore, it is difficult to compile lists of outlets on an item-by-item basis because of merchandising patterns. For example, department stores sell a tremendous variety of items, including clothing, appliances, and furniture; tire and automobile accessory stores also well appliances and toys. Hence compromises must be made in developing a sample of outlets.
(2) The sampling problems are quite different for different items. We can obtain the price rate for electricity by merely visiting the local utility company or the price of newspapers by calling the local publishers. On the other hand, the price of items such as meats and fresh produce vary widely from one grocery store to another, and a fairly large sample of price quotations are required in order to determine the average price of a grocery item with any degree of precision.
(3) There is no clearcut way of identifying the particular outlets that are patronized by the population of consuming families to which the CPI is supposed to apply. The Bureau of Labor Statistics recognizes that the shopping patterns of consuming units now range over a wide geographic area, even though the residences may be confined to a city. Thus the final sample attempts to give proper representation to the downtown and neighborhood areas of the central cities, as well as to suburban areas where so many shopping centers have been developed in recent years.

The field operations of monthly pricing are intricate. Not only must a large field staff be trained and supervised, but one must also obtain the cooperation of store managers, and make provision for businesses that cease operations or come into existence during the ten year period that ordinarily elapses between major revisions. The magnitude of these endeavors is indicated by the fact that food prices are collected in 1775 stores each month and that the Bureau is in touch with about 16,000 outlets for the pricing of nonfood commodities and services.

There is a final sampling problem, not always recognized as such, associated with the CPI. Although it is published monthly, it does not refer to any definite date during the month. The pricing operation has to be almost continual, and it is therefore necessary to choose a sample of points in time at which the prices are obtained. This sampling is not carried out in as formal a manner as are the other sampling operations. Nevertheless, every attempt is made to ensure, for example, that sale and nonsale days for food are represented in their proper proportions, and that a similar balance is maintained for other items, such as newspapers and theater admissions, whose prices may change periodically.

One important goal in statistical design and analysis is to have an objective measure of the precision of sample analyses. Although this goal has not been fully realized in the complex setting of the CPI, substantial progress was made in this direction during the 1959–64 revision, and crude estimates of sampling error have been obtained. Since 1967, these estimates are given in *The Consumer Price Index,* a monthly bulletin published by the Bureau of Labor Statistics. A recent issue of this bulletin states that ". . . any particular (month-to-month) change (in the CPI) of 0.1 percent may or may not be significant. On the other hand, a published change of 0.2 percent is almost always significant, regardless of the time period to which it relates." All indications are that the sampling operations are reasonably well under control, and that uncertainty in the value of the index due to sampling is relatively small compared to the uncertainties arising from other aspects of the process, for example, the effects of quality changes on the index.

SUMMARY

The production of monthly values of a Consumer Price Index by the U.S. Bureau of Labor Satistics is a highly complex undertaking that involves problems of *basic economic theory* (e.g., choice between a price index or a constant-satisfaction index), *measurement and quantification* (e.g., of changes in the quality of items purchased by consumers), *sampling statistics* (definition of, and selection of samples from, a wide variety of populations), and *operations* (e.g., training and supervision of price reporters).

The Index is concerned with a *population of consuming families* and with the *population of cities* in which these families live. A sample of cities serves two purposes. First, from within the selected cities a sample of consumers can be chosen from which it is possible to determine average expenditure patterns. Second, prices are collected in the selected cities to determine the value of the current CPI. The Consumer Expenditure Surveys also define a *population of goods and services* for which consuming families spend their income. From this population a sample must be taken for current pricing purposes. Within each of the index cities there exist *populations of outlets* at which items can be purchased and samples must be chosen to represent these populations. Finally, there is a *population of times* within a month at which price quotations can be obtained, and this population also must be sampled.

It is certainly true that no one is completely satisfied with the CPI in its present form. Improvements can and probably will be made in many parts of the Index, for example, in the basic data on consumer expenditure patterns, in the sampling of outlets and in the collection of price data from these outlets, in the preparation of indexes for a wider variety of subpopulations, and in techniques for handling quality change problems. There will be continuing pressure also to move the CPI more in the direction of a constant-utility type of index. It may even happen that some completely new approach to the construction of the index may be developed, possibly through the use of newer mathematical techniques. No matter what its form, however, the CPI undoubtedly will remain one of the main indicators of the state of the U.S. economy.

PROBLEMS

1. Refer to Figure 1. In 1965 how much would the average urban family have paid for the same amount of goods and services that could have been obtained for $10.00 in 1957–1959?

2. Explain the difference between a "constant-utility" and a "fixed-market-basket" type of index. Which type would you prefer the Bureau of Labor Statistics to use?

3. What is the purpose of dividing up cities into homogeneous groups before choosing the cities to be included in the price survey?

4. Why are all 12 of the largest cities automatically included in the price survey?

5. Suppose the market basket used in computing the CPI in January 1968 cost $750. How much more would it have cost in February 1968 before you concluded that a statistically significant price increase had occurred?

6. Suppose a group of rural congressmen pushed through a bill requiring the Department of Agriculture to find out the cost of living for farm families.
   (a) Could the Department refer the Congressmen to the CPI?
   (b) If not, what steps would it have to take to implement this law?

7. Suppose an anthropologist decided to construct a CPI for a remote community of Northern Albertan gold prospectors which he was studying. In January 1975 he took a survey and found that the prospectors spent on the average:
   (a) 100 ounces of gold for flour, which cost 4 ounces a sack;
   (b) 20 ounces for burros, which cost 100 ounces on the average;
   (c) 40 ounces of gold for whisky, which costs 5 ounces per pint.

In January 1976, the anthropologist finds that flour has gone up to 6 ounces of gold per sack; burros sell for only 80 ounces on the average; and whisky costs 10 ounces a pint.

What is the CPI for January 1976 (January 1975 = 100)?

REFERENCES

Ethel D. Hoover. 1968. "Index Numbers: II. Practical Applications." David Sills, ed., *International Encyclopedia of the Social Sciences,* vol. 7, pp. 159–165. New York: Macmillan and Free Press.

Philip J. McCarthy. 1968. "Index Numbers: III. Sampling." David Sills, ed., *International Encyclopedia of the Social Sciences,* vol. 7, pp. 165–169. New York: Macmillan and Free Press.

Erik Ruist. 1968. "Index Numbers: I. Theoretical Aspects." David Sills, ed., *International Encyclopedia of the Social Sciences,* vol. 7, pp. 154–159. New York: Macmillan and Free Press.

U.S. Department of Labor. *The Consumer Price Index: History and Techniques.* Bulletin No. 1517. Washington: U.S. Government Printing Office.

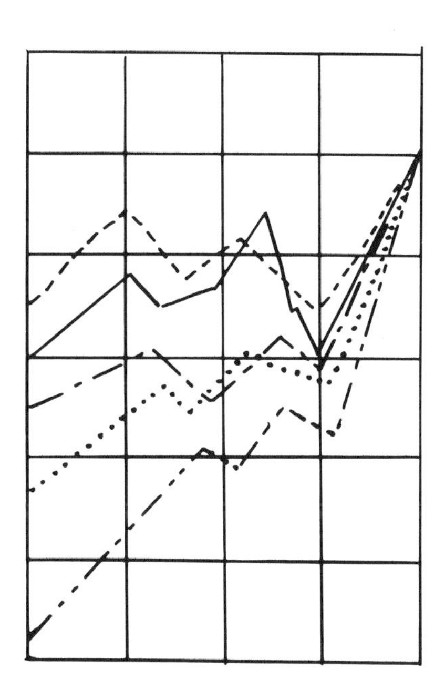

EARLY WARNING SIGNALS FOR THE ECONOMY

Geoffrey H. Moore *Bureau of Labor Statistics*

Julius Shiskin *Office of Management and Budget*

During the sharp business setback of 1937–38, the National Bureau of Economic Research (NBER) was asked by the Secretary of the Treasury, Henry Morgenthau, to devise a system that would signal when the setback was nearing an end.

At that time, quantitative analysis of economic performance in the U.S. was still in a rudimentary stage. The Government's National Income and Product Accounts, today the foundation of so much economic analysis, were just being established. A number of statistical series, including unemployment rates, were being developed or refined by public agencies trying to provide information that would be useful in fighting the Depression. Among business, labor, and academic groups, there was a surge of interest in obtaining better economic intelligence as a guide to policy making.

Under the leadership of Wesley C. Mitchell and Arthur F. Burns, the private, nonprofit NBER had since the twenties assembled and analyzed a vast amount of monthly, quarterly, and annual economic data on prices, employment, production, and so on as a basis for the NBER's major research effort to gain a better understanding of business cycles. From these statistical series, it selected 21 that, on the basis of past performance (dating as far back as 1854), promised to be fairly reliable indicators of business revival. This list was made available to the U.S. Treasury late in 1937 in response to its request and published in May 1938.

This, in brief, is the origin of the statistical indicators so widely used today in analyzing the economic situation, determining what factors are favorable or unfavorable, and forecasting short-run developments.

Since then, the availability, study, and use of economic indicators has been vastly expanded under the leadership of the National Bureau and other public and private agencies.

Especially during a period of change in business activity the indicators, and analyses based on them, are front page news. A drop or rise in the number of persons unemployed, a change in the rate of the nation's total output, a movement up or down in the indexes of consumer or wholesale prices, a rise or fall in new orders for goods—all of these command nationwide attention. Among professional economists interested in the future performance of the economy, indicators have become a major tool of economic forecasting. In this essay, we shall discuss what these statistical measures are and how they are used.

THE INDICATORS IN BRIEF

The indicators have been classified into three groups: those that provide advance warning of probable changes in economic activity, the *leading* indicators; those that reflect the current performance of the economy, the *coincident* indicators; and those that confirm changes previously signaled, the *lagging* indicators.

Coincident Indicators. The most familiar type of indicator, the coincident, measures current economic performance. Gross National Product, industrial production, personal income, employment, unemployment, wholesale prices, and retail sales are examples. Comprehensive in coverage, these indicators show how well the economy is faring in certain important respects. Their movements coincide roughly with, and indeed provide a measure of, aggregate economic activity. They tell us whether the economy is currently experiencing a recession or a slowdown, a boom or an inflation.

A number of government agencies cooperate in the gathering and compilation of these statistics. To produce the Wholesale Price Index, for example, the Bureau of Labor Statistics gathers by mail questionnaire each month about 7000 price quotations on more than 2000 commodities covering the output of manufacturing, agriculture, forestry, fishing, mining, gas and electricity, and public utility industries.

The Census Bureau conducts a monthly survey of 50,000 households to obtain information on employment and unemployment. This survey provides such indicators as the unemployment rate and the total number of persons employed. Indicators of the number of employees on nonagricultural payrolls and total number of man-hours worked are obtained from reports by 160,000 business establishments to the Bureau of Labor Statistics.

The Office of Business Economics of the Department of Commerce is the keeper of a number of statistical series of current economic performance. The most significant is the Gross National Product, which is the total of all goods and services produced by the economy, reported quarterly both in current dollars and adjusted for changes in the price level.

The Federal Reserve Board compiles the Index of Industrial Production, another important coincident indicator.

Leading Indicators. In view of the great impact that economic developments have upon our daily lives and upon long-term economic progress, there is intense interest in indicators that signal in advance a change in the basic pattern of economic performance. Examples are new orders for durable goods, construction contracts, formation of new business enterprises, hiring rates, and the average length of the workweek. These indicators move ahead of turns in the business cycle, primarily because decisions to expand or curtail output take time to work out their effects, while the factors that govern these decisions also take time to produce their influences. The early warning signals provided by leading indicators help to forecast short-term trends in the coincident series and help policy makers to take timely steps to avert, or at least to moderate, unfavorable economic trends.

Many of the leading indicators are produced by the same agencies or surveys that yield the coincident indicators. Thus, the average workweek of production workers, the hiring rate, and layoff rate are figures from the Bureau of Labor Statistics employment survey. The value of manufacturers' new orders for durable goods is compiled by the Census Bureau from the same survey that produces monthly statistics on inventories and shipments.

A number of leading indicators are compiled by private firms. For example, the index of the total value of construction contracts is prepared by the McGraw-Hill Information Systems Co. The number of new business incorporations is provided by Dun and Bradstreet, Inc., and the index of stock prices by Standard and Poor's Corp.

Leading indicators are used more and more widely by business, government, and academic economists in analyzing and forecasting business conditions. In a recent survey by the American Statistical Association, 261 forecasters were asked what principal methods they used, and 103 checked "lead indicators." One of the information sources on which business analysts depend is *Business Conditions Digest* (BCD), a monthly publication of the U.S. Department of Commerce that charts leading and other indicators.

The use of leading indicators has spread to individual states and to foreign countries. New Jersey is one of several states issuing monthly reports on the current position of leading indicators for the state. The Canadian and Japanese governments each have developed reports similar to BCD, and studies of leading indicators have been made for Great Britain, Germany, France, Italy, and Australia. As a result of this widening interest, we are more and more likely to read in newspapers, magazines, and business advisory publications about developments in the leading indicators.

Lagging Indicators and Others. Still another type of indicator is described as lagging. The fluctuations of these indicators usually follow, rather than lead, those of the coincident indicators. Examples are labor cost per unit of output, long-term unemployment, and the yield on mortgage loans.

Finally, there are important economic activities that have not behaved in a sufficiently consistent manner to be appropriately classified as leading, coincident, or lagging, but are nevertheless relevant to an overall appraisal of the current situation of the economy and prospective trends. Examples are government expenditures and the balance of payments.

CHARACTERISTICS OF INDICATORS

The selection of indicators has been guided by two considerations. First, does the measured process play a significant role in a widely accepted explanation of short-term economic fluctuations? While the recurrence of successive waves of rapid growth and slower growth or decline in business activity is generally acknowledged, many different explanations of the underlying causes have been advanced. Some economists lay primary stress on the relations between investments in inventory and fixed capital, on the one hand, and final demand, on the other (John Maynard Keynes, Paul Samuelson). Others assign a central role to the supply of money and credit (Milton Friedman). Still others look for clues in the relationships among prices, costs, and profits (W. C. Mitchell, A. F. Burns). All these factors undoubtedly influence the course of business activity, and some may be more important at one time than another, but there is no general agreement on which is the most important. Hence it is prudent to consider a variety of indicators that reflect all the processes.

The second consideration in selecting indicators has been their empirical record. How closely correlated are the fluctuations in a given series with those in aggregate economic activity? How consistent has been the timing record of a given series compared to aggregate economic activity; that is, does it consistently lead, coincide with, or lag aggregate economic activity? Specified criteria have been applied to the hundreds of economic series from which a list of indicators could be selected. These pertain to the economic significance of the series, its historical record, and various properties affecting its reliability as a current statistic.

In recent years, most of the research and testing in this field has been carried on by the National Bureau of Economic Research. Since publishing its first list in 1938, the Bureau has revised the list in 1950, 1960, and 1966. The most recent list appears in a book by the authors of this essay, published by the NBER in March 1967, under the title *Indicators of Business Expansions and Contractions.* Many of these indicators have been used for years in appraising economic conditions, but 13 were incorporated for the first time in the latest list. Among these new indicators are job openings at U.S. Employment Service offices, delinquency rates on installment loans, export orders for durable goods, and man-hours of nonagricultural employment. Up-to-date figures on these indicators and other related series are published each month in *Business Conditions Digest.*

An understanding of the role of the selected indicators in initiating or reacting to short-term economic fluctuations is aided by the two principles of classification used in presenting the current list of 88 indicators. First is a grouping by *cyclical timing,* as explained above, with 36 leading, 25 coincident, and 11 lagging indicators (16 are not classifiable by timing). The second type of classification is by *economic process,* in which series that pertain to different stages or aspects of the same process are grouped together.

All series are cross-classified according to these two principles. Thus, the cross-reference system shows, under the employment and unemployment category, five leading series representing marginal employment adjustments such as the hiring rate and the workweek (which varies with the amount of overtime worked), eight coincident indicators representing existing job vacancies and comprehensive measures of employment and unemployment, and one lagging indicator representing long-term unemployment. The capital investment category includes ten leading indicators representing the formation of business enterprises and new investment commitments, two coincident indicators representing the backlog of capital investment, and two lagging indicators representing current investment expenditures. Other economic process categories include production, income, consumption and trade; inventories and inventory investment; prices, costs and profits; money and credit; foreign trade and payments; and federal government activity. Together they provide a rather comprehensive view of the economic system.

Another innovation of the National Bureau publication is a method of assigning numerical scores, or *weights,* to each indicator, ranging from 0 to 100. The scoring plan covers six major elements: economic significance, statistical adequacy, historical conformity to business cycles, cyclical timing record, smoothness, and promptness of publication. The ratings throw into clearer perspective the characteristic behavior and limitations of each indicator, aid in their classification and incidentally suggest ways to improve them for purposes of short-term forecasting (see Table 1). For example, the low score of series 31 (change in inventories) for smoothness (column 6) is a warning that sharp erratic movements from month to month are to be expected and that it may take several months to discern a new trend.

The short, substantially unduplicated list of 26 principal indicators shown in Table 1 provides a convenient way of summarizing the current situation and outlook. This list includes 12 leading, 8 roughly coincident, and 6 lagging indicators. The series selected for each of these groups have high scores and cover a broad range of economic processes.

SUMMARY INDEXES

In studying the current economic situation, we can proceed by examining numerous and varied aspects of the economy so as to be sure that all the relevant points are covered or by examining just a few selected indicators that make it easy to grasp the overall trends. Most business analysts move back and forth from a detailed examination of many sectors to a broad view of the overall situation, and there is a feedback of information and insight from one view to the other. The indicator scheme just described makes it appropriate to ask whether a broader type of summary, combining indicators with similar short-term timing behavior, but pertaining to different aspects of activity, can be constructed.

The National Bureau studies make it possible to do this; that is, series that usually lead in the business cycle can be combined into one index, coincident series into another, and lagging series into a third. Similarly, within the leading group alone, series that represent orders or commitments for capital investment projects can be combined into one index, those representing inventory investment or materials purchasing into a second, and those representing sensitive flows of money or credit into a third. A suitable set of weights to be used in combining the series is provided by the scores referred to earlier.

In the sense that they are not expressed in a common unit such as dollars or tons, the series selected for inclusion in each of the indexes are heterogeneous. In the special sense that they measure related aspects of business change, are sensitive to business cycles, and experience similar timing behavior during cyclical fluctuations, they are homogeneous, however. In this respect, some of the best known aggregates are heterogeneous. For example, Gross National Product includes the change in inventories, a leading indicator; consumption expenditures, a coincident indicator; and investment expenditures, a lagging indicator.

TABLE 1. Scores for 26 Economic Indicators on 1966 NBER Short List

CLASSIFICATION AND SERIES TITLE	Average Score*	SCORES, SIX CRITERIA*					
		Economic Significance	Statistical Adequacy	Conformity	Timing	Smoothness	Currency
Leading indicators (12 series)							
1. Avg. workweek, prod. workers, mfg.	66	50	65	81	66	60	80
4. Nonagrl. placements	68	75	63	63	58	80	80
12. Index of net business formation	68	75	58	81	67	80	40
6. New orders, dur. goods indus.	78	75	72	88	84	60	80
10. Contracts and orders, plant and equip.	64	75	63	92	50	40	40
29. New building permits, private housing units	67	50	60	76	80	60	80
31. Change in book value, mfg. and trade inventories	65	75	67	77	78	20	40
23. Industrial materials prices	67	50	72	79	44	80	100
19. Stock prices, 500 common stocks	81	75	74	77	87	80	100
16. Corporate profits after taxes, Q	68	75	70	79	76	60	25
17. Ratio, price to unit labor cost, mfg.	69	50	67	84	72	60	80
113. Change in consumer instal. debt	63	50	79	77	60	60	40
Roughly Coincident Indicators (8 series)							
41. Employees in nonagrl. establishments	81	75	61	90	87	100	80
43. Unemployment rate, total (inverted)	75	75	63	96	60	80	80
200. GNP in current dollars, Q	80	75	75	92	82	100	50
205. GNP in constant dollars, Q	73	75	75	91	58	80	50
47. Industrial production	72	75	63	94	38	100	80
52. Personal income	74	75	73	89	43	100	80
56. Mfg. and trade sales	71	75	68	70	80	80	40
54. Sales of retail stores	69	75	77	89	12	80	100
Lagging indicators (6 series)							
44. Unempl. rate, persons unempl. 15+ weeks (inverted)	69	50	63	98	52	80	80
61. Bus. expend., new plant and equip., Q	86	75	77	96	94	100	80
71. Book value, mfg. and trade inventories	71	75	67	75	66	100	40
62. Labor cost per unit of output, mfg.	68	50	70	83	56	80	80
72. Comm. and indus. loans outstanding	57	50	47	67	20	100	100
67. Bank rates on short-term bus. loans, Q	57	50	55	82	47	80	25

Source: "Indicators of Business Expansions and Contractions," 1967, New York: National Bureau of Economic Research, Inc.

* The scores run a scale from 0 to 100. For example, a series with a random relation to business cycles would be expected to score 0 for conformity and timing (columns 4 and 5). The average score is an average of the six criteria scores with smoothness and currency weighted one-half each.

The procedure used in constructing the indexes allows for the fact that some indicators, such as new orders, typically move in wide swings while others, such as the average workweek, experience narrow (but, nevertheless, significant) fluctuations. Each indicator is adjusted in such a way that, apart from its weight, it has the same opportunity to influence the index as any other indicator. The indexes themselves are adjusted in a similar manner, with the result that their swings are of the same order of magnitude on the average (namely, 1.0% per month) and can be compared readily. For example, if the most recent monthly increase in an index is 2.0, it is rising twice as fast as its average rate of change in the past; if the increase is 0.5, it is rising only half as fast as the historical average.

The index for the leading group is also subject to a further adjustment, designed to make its long-run trend the same as that of the index of coincident series. The major difference that remains is in cyclical timing, with the leading index typically moving first, the coincident index next, and the lagging index last, as can be seen in Figure 1, a type of chart currently published in *Business Conditions Digest*.

It is noteworthy that when the scoring system used to provide weights for the individual indicators is applied to the indexes themselves, the scores earned by the indexes are higher than those for any of the component indicators. In this sense, the indexes are a superior form of indicator. But to understand and interpret their movements, close study of the components is essential.

HOW THE INDICATORS HAVE PERFORMED

One of the uses to which the leading index can be put is to make explicit forecasts for short periods ahead of GNP, total employment, or other variables. Some tests of a simple method for doing this have been made, with promising results. For instance, when data for the third quarter become available, say in October, the percentage change in GNP (in current dollars) between the current calendar year and the next can be forecast by observing the percentage change in the leading index between the third quarter and the *preceding* fiscal year.

Annual forecasts obtained in this manner compare favorably in accuracy with what most forecasters have been able to achieve. During 1962–68, the average error in the GNP forecasts contained in the *Economic Report of the President* was 1.3 percentage points, while during this period the leading index method would have produced an average error of about 1.0 percentage points. In 1968, this superiority was not maintained because the forecast of increase in GNP obtained from the leading index was about 6%, the *Economic Report* (February 1968) put it at nearly 8%, but the actual change turned out to be 9%. For 1969, the leading index (using data through September

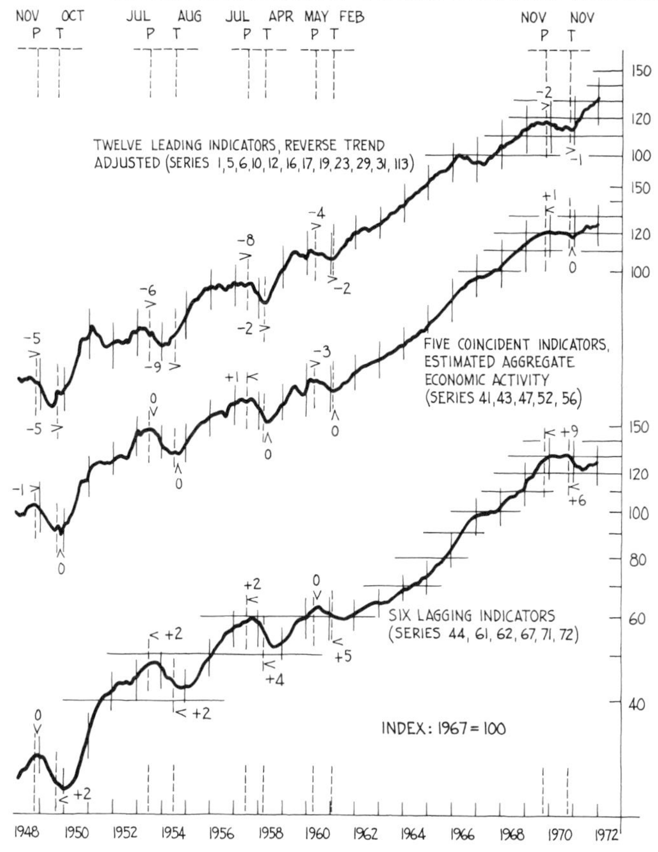

FIGURE 1

Composite indexes. Dates at top of chart are reference peak and trough dates when expansion ended and recession began (P) and when recession ended and expansion began (T). Numbers on chart are length of lead (—) or lag (+), in months, from index peak or trough to reference peak or trough. Source: U.S. Bureau of the Census

1968) forecast an increase in GNP in the neighborhood of 7%, significantly lower than the actual increase in 1968. The *Economic Report* (January 1969) also forecast 7%. The actual increase was about 7.5%. The leading index forecasts can be revised monthly as additional data become available.

The method clearly is no substitute for a carefully reasoned approach to the economic outlook. It merely helps to summarize the information contained in the leading indicators regarding the near-term future course of GNP or other variables that are systematically related to the business cycle. Hence it provides the forecaster with information useful in developing his actual forecast. It can be used also as a standard against which to judge past efforts and might be of assistance in improving upon them, but it is not a recipe that will tell the forecaster everything he should know.

SUMMING UP

A few concluding observations may be helpful to those who would use these indicators as an aid in interpreting current trends.

(1) The change in pace of the leading indicators foreshadows increasing strength or weakness in aggregate economic activity. Sometimes weakness in the leading indicators is followed by a recession, that is, an extended, substantial, and broadly diffused decline in aggregate economic activity, as in 1957–58. At other times, however, a decline in the leaders is followed only by a slowdown in the rate of expansion of aggregate economic activity, as in 1962 and 1967 (see Figure 1). Indeed, if the response of government policy to a decline in the leading indicators is prompt and vigorous enough, there may be no unfavorable developments in aggregate activity at all.

(2) Whether in any particular instance a decline in the leading indicators signals a retardation or a recession will eventually be determined by the movements of the coincident indicators, that is, by such measures of aggregate economic activity as total production, employment, income, consumption, trade and the flow of funds. In view of the fact that economic upswings and downswings have several relevant dimensions (length of swing, how deep it swings, and how pervasive the effect), and that the data themselves are fallible and subject to revision, several months may elapse before a current decline can be reliably appraised.

(3) The lagging indicators should not be neglected. They may provide evidence confirming a change in trend that has appeared earlier in the leading or coincident indicators. Their value in this role is all the greater because the factors that make them lag also make them relatively impervious to erratic movements. In addition, some of the lagging indicators have an important bearing on the *subsequent* behavior of the leading indicators. For example, when unit labor costs, a lagging indicator, rise rapidly, they may exert a downward pressure on profit margins, one of the leading indicators. In this connection, it is often the relation between the movements in lagging and coincident indicators (in this case costs and prices) rather than those of lagging indicators alone that is crucial.

(4) The ultimate objective of the work with indicators is the *prevention* of unfavorable developments, especially recession and inflation. The intent is not to compile good forecasting records; it is rather to develop warning signals which come sufficiently early to assure that effective preventive measures can be taken in time. Hence a successful record of forecasting recession and inflation would attest the failure of economic policy, while successful economic policy might well relegate many accurate forecasts to the limbo of apparent failure.

What can be said of the usefulness of business cycle indicators on balance? It seems clear from the record that the business indicators are helpful in judging the tone of current business and short-term prospects. But because of their limitations, the indicators must be used together with other data and with full awareness of the background of business and consumer confidence and expectations, governmental policies, and international events. We also must anticipate that the indicators will often be difficult to interpret, that interpretations will sometimes vary among analysts, and that the signals they give will not always be correctly interpreted.

Indicators provide a sensitive and revealing picture of the ebb and flow of economic tides that a skillful analyst of the economic, political, and international scene can use to improve his chances of making a valid forecast of short-run economic trends. If the analyst is aware of their limitations and alert to the world around him, he will find the indicators useful guideposts for taking stock of the economy and its needs.

PROBLEMS

1. From Table 1 find the leading indicator with the highest average score of all leading indicators. Do a similar thing for coincident and lagging indicators.

2. Explain the difference between leading, lagging and coincident indicators.

3. If you graph indicators from month to month, which type would tend to have the smoothest graphs—leading, coincident, or lagging indicators?

4. How do economists use the economic indicators described?

5. If you had a new index of economic activity how would you decide whether it was a leading, lagging or coincident indicator, or whether it could not be classified as any of these?

6. Have the indicators been changed since this article was written?

7. The Wholesale Price Index (WPI) comes out each month. Suppose we compute a new index—call it the Average Recent Wholesale Price Index

(ARWPI)—by averaging the values of WPI for the 6 previous months (so, for example, ARWPI for December = $^1/_6$[(WPI for Dec.) + (WPI for Nov.) + (WPI for Oct.) + (WPI for Sept.) + (WPI for Aug.) + (WPI for July)]).

(a) Will ARWPI be a lagging, leading or coincident indicator? Why?

(b) Will ARWPI be smoother or less smooth than WPI?

8. (a) If you wanted to know whether the business cycle was in a "down" phase or an "up" phase in the early 1950's, would you rather see unemployment figures or stock prices for this period?

(b) If you wanted to know in which months of the 1950's the business cycle "changed directions," would you rather see unemployment figures or stock prices?

HOW ACCOUNTANTS SAVE MONEY BY SAMPLING

John Neter *University of Minnesota*

Accountancy and statistics are regarded by many people as two of the dullest subjects on earth. The essays in this volume, it is hoped, will change people's views about statistics. This essay deals with important uses of statistics in accounting practice, and it may also reveal some interesting facets of accounting.

All of us, after all, want to use our money efficiently and effectively. We shall see how the use of statistical sampling in accounting saves money for railroads and airlines as they face problems of dividing revenues among several carriers. Similar statistical sampling methods are used in other areas of accounting and auditing work. Indeed they are used in many fields of business, government, and science.

Accountants and auditors traditionally have insisted on accuracy in the accounting records of firms and other organizations. This insistence has led them to do much work on a complete, 100% basis. For instance, an auditor may want to check the value of the inventory that a firm has on hand. To do this, he may examine the entire inventory; that is, he may actually count how many units of each type of inventory item are on hand, determine the value of each kind of unit, and thus finally obtain the total value of the inventory.

As another instance, an auditor may want to know the proportion of accounts receivable that have been owed for 60 or more days. This information may be needed to verify a reserve for bad debts. Accounts that have not been paid within 60 days are more susceptible to bad debt losses than accounts that have been open a shorter time. In order to establish the proportion of accounts receivable that are 60 days old or older, the auditor may examine every single account receivable held by the firm and determine for each the amount of money owed for 60 or more days.

Is it necessary to conduct these 100% examinations of inventory, of accounts receivable, or of similar collections, in order to obtain the figures that the accountant needs? More specifically, could a sample adequately provide the information needed by the accountant without all of the tedious work necessary for a complete, 100% examination? Let us focus on the inventory items.

In statistical terminology, the group of inventory items for which the total value is to be ascertained is called the *population* of interest. A *sample* selected from such a population consists of some, but not all, of the items in the population. A sample is selected to find out about characteristics of the population without looking at every element of the population.

The cost of examining a relatively small sample of inventory items is usually less than the cost of a complete examination because the sample requires an examination of fewer items. But are the results based on the relatively small sample almost as good as those from a complete examination?

Experience with sampling in many areas has shown that relatively small samples frequently provide results that are almost as good as those obtained from a complete examination, while at the same time the sample results cost considerably less. Indeed, sometimes the sampling results are even better than those from a 100% examination. That statement may seem startling, but consider the task of taking an inventory in a large company. Many persons are required for the task. Because of the size of the undertaking, it may be hard to give thorough training to these persons, and the quality check on the work may have to be limited. On the other hand, a small sample of the inventory items would require fewer persons, and therefore they could be trained better. Furthermore, the quality control program for the inventory could be more rigorous when a smaller number of persons are involved. The net effect might well be that the sample results are more accurate than the 100% enumeration! That is, the gains in accuracy from better training and quality control with a small sample may more than balance the sampling error introduced by selecting only a sample of inventory items instead of all of them. Of course, the sampling must be done intelligently and properly. The study of sampling is an important part of statistics.

THE CHESAPEAKE AND OHIO FREIGHT STUDY

Statements that relatively small samples can provide results almost as good as those from a complete examination, or indeed sometimes even better, are often not convincing by themselves. Statisticians have therefore often found it helpful to conduct studies that compare the results of a sample with those of a complete enumeration. Such a study was made by the Chesapeake and Ohio Railroad Company in determining the amount of revenue due them on interline, less-than-carload, freight shipments. If a shipment travels over several railroads, the total freight charge is divided among them. The computations necessary to determine each railroad's revenue are cumbersome and expensive. Hence, the Chesapeake and Ohio experimented to determine if the division of total revenue among several railroads could be made accurately on the basis of a sample and at a substantial saving in clerical expense.

In one of these experiments, they studied the division of revenue for all less-than-carload freight shipments traveling over the Pere Marquette district of the Chesapeake and Ohio and another railroad (to be called A for confidentiality), during a six-month period. The waybills of these shipments constituted the population under examination. A waybill, a document issued with every shipment of freight, gives details about the goods, route, and charges. From it, the amounts due each railroad can be computed. The total number of waybills in the population was known, as well as the total freight revenue accounted for by the population of waybills. The problem was to determine how much of this total revenue belonged to the Chesapeake and Ohio.

For the six-month period under study, there were nearly 23,000 waybills in the population. Since the amounts of the freight charges on these waybills vary greatly (some freight charges were as low as $2.00, others as high as $200), it was decided to use a sampling procedure which is called *stratified sampling*. With this procedure, the waybills in the population are first divided into relatively homogeneous subgroups called strata. The subgroups in this instance were set up according to the amount of the total freight charge, since the amount due the Chesapeake and Ohio on a waybill tended to be related to the total amount of the waybill. That is, the larger the total amount of a waybill, the larger tended to be the amount due the Chesapeake and Ohio on that waybill. Specifically, the strata were as follows:

Stratum	Waybills with Charges Between
1	\$ 0 and \$ 5.00
2	\$ 5.01 and \$10.00
3	\$10.01 and \$20.00
4	\$20.01 and \$40.00
5	\$40.01 and over

Note that each stratum contains waybills with total freight charges of roughly the same order of magnitude. Because of the general tendency by which the amount due the Chesapeake and Ohio varied with the total freight charge on a waybill, each stratum is relatively more homogeneous with respect to the amount of freight charges due the Chesapeake and Ohio. At the same time, the strata differ substantially from one another.

Statistical theory then was used to decide how large a sample from each stratum must be selected so that the amount of the revenue due the Chesapeake and Ohio could be estimated with required precision from as small a sample as possible. One piece of information needed for this determination is the number of waybills in each stratum. The sampling rates decided on for the strata were:

Stratum	Proportion to Be Sampled
1	1%
2	10%
3	20%
4	50%
5	100%

Note that this theory led to larger sampling rates in the strata containing wider ranges of freight charges and smaller sampling rates in the strata containing narrow ranges of freight charges. To understand this, consider stratum 1, containing waybills with charges between \$0 and \$5.00. Here the variation between the waybill amounts is small, and therefore a small sample will provide adequate information about the amounts of all of the waybills in that stratum. On the other hand, stratum 4, containing waybills with charges between \$20.01 and \$40.00, has much greater variation. A larger sample is therefore required in this stratum to obtain adequate information about the amounts of all waybills in that stratum. In an unreal extreme situation with all the waybills in a stratum having the same amount due the Chesapeake and Ohio, a sample of just one waybill would provide all the information about the waybill amounts in that stratum.

Once the sample sizes were determined, the next problem was to select the samples from each stratum. For a statistician to be able to evaluate the precision of the sample results, that is, how close the sample results are likely to be to the relevant population characteristic, the sample must be selected according to a known probability mechanism. Various methods of probability sampling are available. One is called *simple random sampling.* This type of sample may be directly selected by use of a table of random numbers, a portion of which is illustrated in Table 1. How might Table 1 be used to select a simple random sample from each of the strata? Consider stratum 1 and suppose it contains 9000 waybills, which we label with four-digit numbers from 0001 to 9000. We want to obtain four-digit numbers from the table; we might start in the upper left-hand corner, using columns 1 through 4. The first number obtained is 1328. Our first sample waybill is then the one numbered 1328. Our second sample waybill would be 2122. The next number from the table of random digits is 9905, but there are only 9000 waybills in the stratum, so we pass over this number and go on to the next one, which is 0019. This process would be continued until the required sample of 90 (1% of 9000) has been obtained. The digits in the table of random digits are generated so that all numbers (four-digit numbers in our case) are equally likely.

Another method of selecting waybills from each stratum is called *serial number sampling,* and this was the method actually used by the Chesapeake and Ohio Railroad. In this procedure, the sample within each stratum is selected according to certain digits in the serial number of the waybill. In this particular case, the last two digits in the serial number of the waybill were used. To explain how these last two digits are used to select the sample, consider stratum 1, with its 1% sample. The number of possible pairs of digits appearing in the last two places of the serial number (00, 01, 02, . . . , 99) is 100. If one of these pairs is chosen from a table of random digits and all waybills with these last two digits in their serial number selected for the sample, it will be found that about 1% of the stratum is included in the sample. For stratum 1, the random number turned out to be 02. Therefore, all waybills with freight charges of $5 or less whose last two serial number digits are 02 were selected for the sample. The serial number digits used for the other strata, as well as the sampling rates, were as follows:

Stratum	Proportion to Be Sampled	Waybills with Numbers Ending in:
1	1%	02
2	10%	2
3	20%	2 or 4
4	50%	00 to 49
5	100%	00 to 99

Table 1. Portion of a Table of Random Digits

LINE	(1)–(5)	(6)–(10)	(11)–(15)	(16)–(20)	(21)–(25)	(26)–(30)	(31)–(35)
101	13284	16834	74151	92027	24670	36665	00770
102	21224	00370	30420	03883	94648	89428	41583
103	99052	47887	81085	64933	66279	80432	65793
104	00199	50993	98603	38452	87890	94624	69721
105	60578	06483	28733	37867	07936	98710	98539
106	91240	18312	17441	01929	18163	69201	31211
107	97458	14229	12063	59611	32249	90466	33216
108	35249	38646	34475	72417	60514	69257	12489
109	38980	46600	11759	11900	46743	27860	77940
110	10750	52745	38749	87365	58959	53731	89295
111	36247	27850	73958	20673	37800	63835	71051
112	70994	66986	99744	72438	01174	42159	11392
113	99638	94702	11463	18148	81386	80431	90628
114	72055	15774	43857	99805	10419	76939	25993
115	24038	65541	85788	55835	38835	59399	13790
116	74976	14631	35908	28221	39470	91548	12854
117	35553	71628	70189	26436	63407	91178	90348
118	35676	12797	51434	82976	42010	26344	92920
119	74815	67523	72985	23183	02446	63594	98924
120	45246	88048	65173	50989	91060	89894	36036

Source: *Table of 105,000 Random Decimal Digits.* Interstate Commerce Commission, Bureau of Transport Economics and Statistics, May 1949.

Since the serial numbers appear prominently on the waybills, this procedure is a simple one for selecting the sample. Furthermore, in this case, experience indicates that it provides essentially the equivalent of a simple random sample from each stratum.

Altogether, 2072 waybills out of 22,984 in the population (9%) were chosen according to this procedure. For each waybill in the sample, the amount of freight revenue due the Chesapeake and Ohio was calculated. For each stratum, the total amount due for the population of waybills was then estimated, and these estimates were added to obtain an estimate of the total amount of freight revenue due the Chesapeake and Ohio on the almost 23,000 waybills in the population. Because this was an experiment, a complete examination of the population was also made, so that the sample result could be compared with the result obtained from an analysis of all waybills in the population. The findings were:

Total amount due Chesapeake and Ohio on basis of complete examination of population	$64,651
Total amount due Chesapeake and Ohio on basis of sample	64,568
Difference	$ 83

Thus, a sample of only about 9% of the waybills provided an estimate of the total revenue due the Chesapeake and Ohio within $83 of the figure obtained from a complete examination of all waybills. Because the sample cost no more than $1000, while the complete examination cost about $5000, the advantages of sampling are apparent. It just does not make sense to spend $4000 to catch an error of $83. Furthermore, although the error in this instance was against the Chesapeake and Ohio, the next time it may be against the other railroad, so that the long run cumulative error is relatively even smaller.

OTHER RAILROAD AND AIRLINE SAMPLING STUDIES

The Chesapeake and Ohio conducted the same type of test for interline passenger receipts. They studied tickets sold during a five-month period to commercial passengers traveling only on the Chesapeake district of the Chesapeake and Ohio and on two other railroads, A and B. The findings are shown in Table 2. Again, these results dramatically demonstrate the ability of relatively small samples to provide precise estimates of the total revenue due the Chesapeake and Ohio.

Airlines also have used statistical sampling to estimate their share of the revenue on tickets for passengers traveling on two or more airlines. Three airlines tested statistical sampling during a four-month period. In that time,

TABLE 2. Results of Passenger Ticket Study

	100% EXAMINATION	5% SAMPLE	DIFFERENCE	
			Dollars	**Percent**
Railroad A				
Total number of tickets	14,109			
Total revenue	$325,600			
Chesapeake and Ohio portion of total revenue	$212,164	$212,063	−$101	−0.05%
Railroad B				
Total number of tickets	7,652			
Total revenue	$128,503			
Chesapeake and Ohio portion of total revenue	$ 79,710	$ 80,057	+$347	+0.44%

the degree of error in the sample estimate based on relatively small samples did not exceed 0.07% (that is, $700 in $1,000,000) for any of the three airlines. On the basis of this experiment, wider use of statistical sampling in settling interline accounts has been made. At one point in time, the sample consisted of about 12% of the interline tickets and the cumulative sampling error was running at less than 0.1%. The clerical savings were estimated to be near $75,000 annually for some of the larger carriers and more than $500,000 for the industry.

Statistical sampling in accounting and auditing has also been used to estimate the value of inventory on hand, the proportion of accounts receivable balances that are 60 days old or older, and the proportion of accounts receivable balances that are acknowledged as correct by the customer. In each instance, it has been demonstrated that a relatively small sample, carefully drawn and examined, can furnish results that are of high quality and at a much lower cost than with a complete examination.

To summarize, statistical sampling consists of the selection of a number of items from a population, with the selection done in such a way that every possible sample from the population has a known probability of being chosen. Frequently, a statistical sample can provide reliable information at much lower cost than a complete examination. Also, a statistical sample often can provide more timely data than a complete enumeration of the population because fewer data have to be collected and smaller amounts of data need to be processed. Finally, a statistical sample can sometimes provide more accurate information than a complete enumeration when quality control over the data collection can be carried on more effectively on a small scale.

PROBLEMS

1. Explain the difference between simple random sampling and serial random sampling.

2. Suppose a university administrator is considering ordering some new desks for classrooms. He needs to find out how many desks already in use need to be replaced.

(a) Should he consider using sampling methods in this situation? What are the arguments for sampling? Against?

(b) If he did use sampling methods, what would the *population* be?

3. Why was stratified sampling used in the C & O freight study?

4. Refer to Table 2. Add the Railroad A and Railroad B ticket revenues, and find the difference in percent between a five percent sample and a one hundred percent examination.

5. In the C & O freight study, how large a percentage of the total amount due C & O was the result of error due to sampling?

6. An army psychologist wants to take a sample of 1000 enlisted men to find out their attitudes towards the "new Army." He obtains a list of 10,000 enlisted men arranged by squads; each squad has ten men, with a sergeant heading the list, then a corporal, followed by eight privates.

(a) Would you recommend that the psychologist use serial number sampling (using the digits 0–9) to choose a sample of 1000 from this list of 10,000 men? Why?

(b) If the psychologist used serial number sampling, what would be the chance of getting only sergeants in his sample? What if he used simple random sampling?

(c) Answer the questions in (a) and (b) if the psychologist used a list which placed the 100 enlisted men in alphabetical order.

7. Suppose C & O and railroad A sampled tickets to determine their share of revenue from interline passenger receipts every month for a year. For how many months would you expect the sampling error to favor C & O?

8. Use Table 1 to draw a random sample of twenty-five two-digit numbers. How many are even? How many have both digits even? Do the same for a random sample of 100 two-digit numbers. Compare your answers to those obtained by the other students in your class. What conclusions can you draw?

REFERENCES

"Can Scientific Sampling Techniques Be Used in Railroad Accounting?" *Railway Age,* June 9, 1952, pp. 61–64.

R. M. Cyert and H. Justin Davidson. 1962. *Statistical Sampling for Accounting Information.* Englewood Cliffs, N. J.: Prentice-Hall.

W. Edwards Deming. 1960. *Sample Design in Business Research.* New York: Wiley.

Henry P. Hill, Joseph L. Roth, and Herbert Arkin. 1962. *Sampling in Auditing.* New York: Ronald.

Elbert T. Magruder. 1955. *Some Sampling Applications in the Chesapeake and Potomac Telephone Companies.* Chesapeake and Potomac Telephone Companies.

John B. O'Hara and Richard C. Clelland. 1964. *Effective Use of Statistics in Accounting and Business.* New York: Holt.

Morris J. Slonim. 1960. *Sampling in a Nutshell.* New York: Simon and Schuster.

Robert M. Trueblood and Richard M. Cyert. 1957. *Sampling Techniques in Accounting.* Englewood Cliffs, N. J.: Prentice-Hall.

Lawrence L. Vance and John Neter. 1956. *Statistical Sampling for Auditors and Accountants.* New York: Wiley.

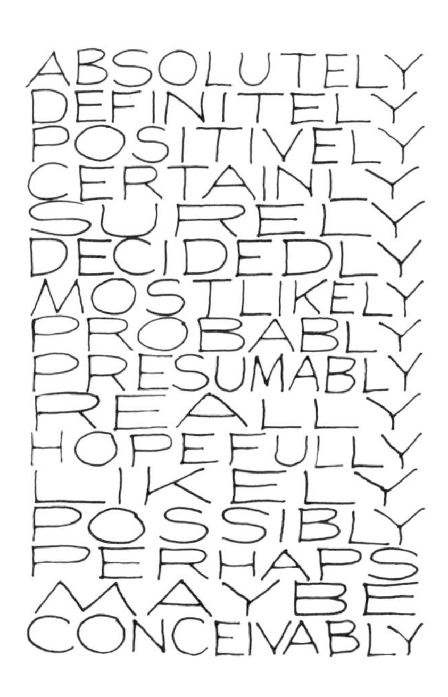

THE USE OF SUBJECTIVE PROBABILITY METHODS IN ESTIMATING DEMAND

Hanns Schwarz *Vice President, Daniel Yankelovich, Inc.*

THE CONTRIBUTIONS of statistics to sampling and analysis of data in marketing research are widespread and liberally documented. For example, we generate facts about the behavior and attitudes of very large populations from studies of relatively small (and inexpensively gathered) samples of respondents. The methods for selecting such small groups to represent large populations come from statistical sampling theory. Such sample surveys are now common practice and indispensable in research. Methods of analyzing data emanating from surveys (as well as from other sources) are also research tools provided by statistics.

Moreover, new statistical applications are being developed continually. Some of these relate neither to survey sampling nor to analysis of data, but to the basic questioning process itself—the essential core of marketing research. In 1971, these applications are not yet in widespread use nor are they well documented. This paper describes one technique which has been used successfully in recent marketing research: the use of subjective probability methods in predicting demand.

THE NEED TO ESTIMATE DEMAND BY SURVEY

One important function of commercial research is to make estimates of consumer demand for a product or service. On occasion, such estimates are required for existing products, although forecasting methods based on past sales data are then usually entirely adequate. When new or modified products or services are involved, however, other techniques are called for. In some cases, test markets are used and sales results in these few markets become the basis for national estimates. Experimental variations of market testing using simulated retail outlets are used in instances in which information is required prior to a commitment to large-scale production, in which a product is not far enough along in production to supply one or more entire test markets, in which secrecy is considered desirable, in which there is not enough time for a test market, or in which the test market method has not proved to be particularly useful in the past.

In many instances neither a test market nor an experimental simulation can be used to predict demand, notably in the case of large, durable products such as cars or major appliances, new services not as yet adapted to the consumer market, or new products only in the concept stage that can neither be sold nor shown except, perhaps, in the form of pictures. In such instances, a research survey of the likelihood of purchase is needed.

THE TROUBLE WITH ESTIMATING DEMAND BY SURVEY

Studies of the likelihood of purchase are probably as old as research itself, and over the years, they have been tried with many variations and with various degrees of technical expertise. But the issue almost always boils down to a question such as "How likely are you to buy_____?" All too often, results of such studies have been impressively wrong, usually ending in estimates of buying intention substantially higher than subsequent levels of actual purchases. In certain cases—notably in attempts to assess interest in a new type of product that does not yet exist—there are instances in which 10% or less of those who said they would buy actually did so when the product reached the market. Of course, the same problem occurs in the case of existing products, too. Research literature abounds with examples of gross discrepancies between stated purchase intentions and subsequent purchasing behavior. A recent example from our own work concerns automobile buying. Of a total of 72 persons who reported that they planned to buy a car within the next six months, only 33 (46%) actually had done so when they were reinterviewed after six months.

The reasons for such overstatement by survey respondents have been explored and discussed many times. The desire of survey respondents to be agreeable—to say "Yes"—particularly because it costs them nothing to do so, has often been cited. Moreover, there is an ego-enhancing aspect to reporting that the purchase of a new car, for example, is imminent. It has been generally agreed that this overstatement phenomenon is not due to conscious falsification but to a failure to consider seriously all of the factors which will play a part when the time of actual purchase comes. In the absence of such serious consideration, it is easiest to say "Yes."

The technique discussed in this paper is based on the experience that most survey respondents are not capable of giving considered answers to direct questions on purchase likelihood. The technique is based on a thorough study and understanding of the particular purchasing process involved, the identification of key factors which are likely to affect this purchase and a quantification of how and to what degree these factors are operating for each survey respondent to affect his likelihood of actual purchase. The method may be termed "a systematic, subjective computation of purchase probability," and it entails the use of subjective probabilities that differ somewhat from classical statistical probabilities. The latter derive either from the full understanding of the mechanics of a process (e.g., dice shooting, coin tossing) or from extensive empirical observation so that they are precise or very nearly precise probabilities. Subjective probabilities on the other hand are not nearly so precise. They are essentially educated guesses derived from judgment, experience, and so on, and generally are used under conditions of uncertainty which very frequently characterize business decisions.

THE USE OF PROBABILITIES IN ESTIMATING DEMAND

In order to demonstrate most clearly how the procedure works, a hypothetical study involving an estimate of consumer demand for a new, atomic-powered cabin cruiser is used for illustrative purposes. While this is a greatly simplified example, it will serve to show how this statistical technique could be used in a variety of subject areas. The essential details of this hypothetical study are that it consists of personal interviews with 200 men who report that they expect to buy a cabin cruiser within the next year. This sample of men is selected through telephone screening to be representative of those "in the market" for a cabin cruiser. These men are shown pictures of the new cruiser as well as a detailed written description and are questioned as to their likelihood of buying it. In addition, answers are solicited to questions relating to the various factors that will be used later to assess overall purchase probability.

TABLE 1. Likelihood of Purchasing Cabin Cruiser as Expressed by Respondents

	NUMBER	PERCENT
Total Respondents	**200**	**100**
Definitely intend to buy	10	5
More likely than not to buy	70	35
Not likely to buy	120	60

The results of the key purchase likelihood question are given in Table 1. Forty percent of the respondents have expressed at least a moderate level of interest in buying the cabin cruiser. Yet it is unreasonable to assume that all of these, in fact, will purchase one. In other words, not every person who has reacted positively has a probability of 1.0 of purchasing it. At this point, the process of making a more realistic assessment of the purchase probability of each respondent starts.

To begin with, the assumption is made that those who have expressed a definite purchase intention are more likely to become actual purchasers than those who have merely admitted to the likelihood of purchase. Therefore, the latter group will be "weighted down" to reflect this lower purchase probability. There is no sure way to determine precisely what weight to apply in the absence of any hard data comparing expressed intention and subsequent purchase of cabin cruisers. Such data have been compiled for other products at other times, however, and can at least serve as a rough guide. For example, in the automobile study cited earlier, the probability of purchase for those who claimed to be planning a purchase was 0.46. Another guide comes from the experience of the researcher and his associates whom he may consult in order to arrive at a weighting scheme based on a "jury of informed opinion." In this hypothetical instance, the consensus is that respondents who have said they are "likely to buy" will receive a weight of 0.4, so that at this intermediate stage, the estimate of demand would be

10 "definitely intend to buy" × weight 1.0 = 10
plus (70 "likely to buy" × weight) 0.4 = 28
divided by 200 = 19% purchasing.

The remainder of the process entails similar weighting of respondents based on various purchase-related factors, thereby further modifying purchase probability. We could argue that further modification may be unnecessary as respondents have considered these purchase factors in answering to the buying intention question. However, there is considerable opinion (Juster 1966) to the effect that responses of this type are, on the whole, not considered ones and that modifying factors do need to be taken into account.

TABLE 2. Weighting the Likelihood that Respondents Will Buy Cabin Cruisers by Time of Intended Purchase

EXPECT TO BUY	WEIGHT
Within 3 months	1.0
In 3 to 6 months	0.8
In 6 to 12 months	0.6

There is likely to be a considerable number of such factors, but for the sake of simplicity, only four will be considered here. Two types of factors exist, the first bearing upon the likelihood of respondent's purchase of *any* cruiser at all within the next year despite his declaration that he will (because this was a requirement for eligibility in this study) and the second bearing upon the likelihood that he will select the particular cruiser being studied.

Of those factors bearing on the purchase of any cruiser, the first concerns just how soon the purchase is expected to be made—within three months, in three to six months or in six to twelve months. Experience indicates that the closer the intended purchase, the more likely that the purchase, in fact, will be made; in other words, those who say they expect to buy within three months are more likely to make actual purchases than those who expect to buy in three to six months, who, in turn, are more likely to become actual purchasers than the six-to-twelve month group. After another round of discussions among several knowledgeable researchers, the weighting scheme shown in Table 2 is developed. In the automotive study cited previously, the data clearly showed that the discrepancy between stated intention and actual purchase was substantially less when the intention was to buy within three months of the time of interview than when the projected purchase data was four or more months away.

The second purchase-related factor deals with the attitudes of the wives of married respondents toward the impending purchase. In this hypothetical instance, the three types of responses coded in the study are shown in Table 3, along with the weights that again are decided by consensus among a group of researchers.

TABLE 3. Weighting the Likelihood that Respondents Will Buy Cabin Cruisers by Wives' Attitudes

RESPONSE	WEIGHT
Wife has no voice in decision	1.0
Wife has a voice and wants to buy a cruiser	1.0
Wife has a voice and prefers alternative purchase	0.6

TABLE 4. Weighting the Likelihood that Respondents Will Buy Cabin Cruisers by Concern over Atomic Power

RESPONSE	WEIGHT
Unconcerned over use of atomic power	1.0
Evidences concern over atomic power	0.4

The third purchase factor bears on the likelihood of purchasing the particular cruiser being studied and deals with the degree of apprehension on the respondent's part about the fact that this is an atomic powered cruiser and might therefore be dangerous in the event of improper shielding of the power source. The weights agreed upon after discussion are shown in Table 4.

The final purchase factor that also bears on the likelihood of selection of the particular cruiser studied concerns the degree to which it fits in with individual price requirements. Those who are planning to spend either more or less than the expected retail price of the cabin cruiser are weighted down as in Table 5.

Using the various weights which have been described, each respondent's probability of purchase is assessed. To illustrate the process:

(1) Respondent A, after seeing a picture of the cabin cruiser and reading a description, reports that he "definitely intends to buy": weight = 1.0.
(2) He claims he will make the purchase within three to six months: weight = 0.8.
(3) He is married but reports that his wife will have no voice in the purchase decision: weight = 1.0.
(4) He reports no apprehension over the use of atomic energy as the power source for this cruiser: weight = 1.0.
(5) The amount he is planning to spend in his upcoming purchase is somewhat less than the expected price of the cruiser: weight = 0.3.

TABLE 5. Weighting the Likelihood that Respondents Will Buy Cabin Cruisers by Consistency with Price Requirements

RESPONSE	WEIGHT
Price in line with plans	1.0
Price higher than planned	0.3
Price lower than planned	0.8

These weights provide the basis for estimating overall purchase probability. Each weight itself is a limited sort of purchase probability—what is called a *marginal probability.* To illustrate how several marginal probabilities are combined to form an overall probability, consider the following illustration.

> The probability of obtaining "heads" in a simple coin toss is 0.5, a marginal probability. The probability of tossing two consecutive heads is 0.25, the product of the marginal probabilities for tossing "heads" on the first toss and for tossing "heads" on the second toss $(0.5) \times (0.5) = 0.25$.

Thus, respondent A's overall purchase probability is simply the product of the various weights (or marginal probabilities): $(1.0) \times (0.8) \times (1.0) \times (1.0) \times (0.3) = 0.24$. The sum of these purchase probabilities across all respondents represents the estimate of demand for the new cabin cruiser. In all, 80 respondents had originally reported that they "definitely intend to buy" or are "likely to buy." If one were to take these responses at face value, he might conclude that as many as 40% (80 of 200) of those in the market may buy this cruiser, though one would be hard pressed to find a modern researcher who would be willing to make so liberal an interpretation of these data. Implicit in this type of estimate is that each respondent who reacts in a positive way is treated as having a purchase probability of 1.0. A more stringent estimating method reduces this purchase probability to 0.4 for those who are only "likely to buy," but leaves the probability at 1.0 for those who report that they will "definitely buy." This, in fact, is how the intermediate estimate of 19% cited earlier was obtained.

Using the system of weighting for various purchase factors brings the estimate lower still; for no one receives a purchase probability of 1.0 without specifically "earning it." In other words, even respondents who will "definitely buy" must give all the "right answers" with respect to the four purchase factors in this illustration before they receive a purchase probability of 1.0. Thus, it would not be at all unlikely that the sum of probabilities for the 80 respondents who reacted positively to the cruiser might be in the area of 5.0 leading to an estimate of demand of 2.5% (5/200), a far cry from 40% and 19%.

In using this subjective probability technique, we must observe one caution. We must overcome the temptation to include an overlong list of factors because if one keeps on multiplying probabilities for too long, the results will come dangerously close to 0. Thus, the factors chosen should be only those that are important purchase influences. In any given case, there are many marginal factors which have only small effects. The analyst must exercise good judgment in limiting the list to the key factors.

Estimates developed through this technique have been accurate. In the study of automobile-buying intentions, for example, the data were highly predictive of actual consumer demand six months later. Although the rule of confidentiality of client data precludes disclosure of specific figures, predicted total sales of the make of automobile as well as various component models were very close to subsequent actual demand.

A PRACTICAL RATIONALE

The technique that has been described here is very unlike the exacting and precise statistical procedures used in sampling and data analysis, described elsewhere in this volume. The development of the weighting schemes appears to be a rather subjective process that is seemingly defenseless against the question: "Why did you choose to use a weight of 0.8 and how can you prove it is better than a weight of 0.6?" It is *not* possible to prove in advance that 0.8 is a better weight than 0.6. However, this lack of certainty should not be permitted to create a "hang-up" to eliminate the use of weights because the means of arriving at them is less than precise. It seems quite clear that the use of weights that are arrived at intelligently is better than having to assume that the person who says "Yes, I will buy," in fact, will do so. In other words, it makes sense to assume that, all other things being equal, the man who is concerned over radiation is less likely to buy an atomic-powered cruiser than one who is not. The method described here makes use of this knowledge on the theory that it would be less accurate to ignore it.

In other respects, too, the method described here is less precise than classical statistical theory would dictate; for example, the assumption of the independence of the various weights is made. Despite some of these loose ends in the theoretical framework of the technique, however, we see only one practical limitation on its usefulness: the skill of the practitioner. Successful application depends on the ability to understand the essential workings of the particular purchasing process being studied and the pinpointing of those key factors that will affect purchasing behavior. As we gain more experience, it is not unlikely that some standardization of weights according to product class may be achieved (e.g., for cars, major appliances, minor appliances) thereby increasing the general usefulness of the method.

On a pragmatic basis, the technique has proved to be a valuable one. It gives good, realistic results and appears to offer a viable solution to the problem of determining purchase likelihood via questionnaire.

PROBLEMS

1. Respondent B after reading about the cruiser and seeing a picture of it is interviewed by the research group. He definitely plans to buy within three months. His wife, however, has a voice in the decision and prefers another type of cruiser. He was planning to spend more than the price of the cruiser. The use of atomic power for propulsion does not bother him.

Use tables two through five to find his marginal probability.

2. Who decides what weights to assign to the purchase factors?

3. Give one reason in favor of subjective probability methods and one opposed to them.

4. Each respondent could be asked to select a weight for himself. For example, a respondent giving a weight of 0.4 would believe that he has a forty percent chance of making a purchase. Comment on this possible method.

5. At the beginning of the school year, a high school guidance counsellor wants to find out how many students in the senior class will go to college next year. 1000 say they plan to go to college, but experience tells him this figure is too high. He decides that two factors are important in shaping the students' final decision: grades and professional plans. He devises a questionnaire and assigns weights as follows.

Group	Grade Point Average	Weight	Group	Professional Plans	Weight
I	3.5–4.0	1	A	"Plan to pursue a definite profession which interests me"	1
II	3.0–3.5	.8	B	"Not sure—will decide by taking courses which interest me"	.8
III	2.0–3.0	.5	C	"Not sure—want to make money somehow"	.4

He submits the questionnaire to the students who say they plan to go to college. He obtains the following result:

PLANS \ GPA	I	II	III
A	150	100	50
B	75	150	175
C	25	100	175

How many seniors does the guidance counsellor estimate will attend college next year?

6. A clothes designer is bringing out a new line of "leisure clothes." He estimates that, of those who claim a "definite interest" in buying these clothes, only 50% of those disagreeing with the statement "the clothes he wears always express a man's *real* personality" will actually purchase; while only 60% of those who say they must wear conventional business suits with ties for work will buy his clothes. The designer figures that 50% × 60% = 30% of those who declare a definite interest in buying, disagree with the statement about clothes and personality, and must wear conventional clothes to work will actually buy his clothes. Is the designer's reasoning correct? Why?

REFERENCE

F. Thomas Juster. 1966. *Consumer Buying Intentions and Purchase Probability.* National Bureau of Economic Research.

PRELIMINARY EVALUATION OF A NEW FOOD PRODUCT

Elisabeth Street *General Foods Corporation*

Mavis B. Carroll *General Foods Corporation*

MANY AMERICANS like to have at hand an easy-to-prepare, nutritious, on-the-run meal. Our company has been developing such a product, called H.

Development of such a product calls for a thorough evaluation. In this essay, we limit ourselves to two aspects of the evaluation, the protein content of H and its tastiness. We shall show how statistics was central in answering our questions.

The palatability of a product can be determined by having people taste the prepared product and evaluate its acceptability both overall and by specific attributes. The nutritive quality of a food product can be determined by feeding it to animals whose metabolic processes are very similar to ours. Both types of tests should be designed, that is, planned in detail in advance. This helps avoid a consistent error in one direction (bias) and ensures that the proper number of people and animals are included in each study to answer with a fair degree of confidence the questions being posed. Some of the steps in the design and analysis of such tests will be described here.

PROTEIN EVALUATION

The protein content of H, in one sense, was satisfactory, as calculations based on the constituents of H showed. We were not certain, however, about the efficiency of the protein in H under conditions of actual use, and in addition, we wanted to compare two forms of H, one solid and the other in liquid form.

A rat-feeding study can give practical support to the high protein claim and a way of comparing the two variants of H because the rats' responses would be affected by any interaction of the ingredients in the formula or by a shortage of an essential amino acid such that the protein would be less efficient. Neither of these conditions would be indicated by the paper calculation.

Previous experience had shown that 28-day feeding of 10 to 15 rats on a diet gives a fairly reliable estimate of the diet's protein efficiency. For this feeding study, as for all such studies, male rats, newly weaned, were used. Male rats grow faster than female rats, and while weanlings, they are in a period of maximum growth rate. Adult rats are not used, for their weights are stabilized, and it is the animals' weight gain that is of primary interest.

Besides comparing the two forms of H, the experiment also compared each with a casein control diet, which served two purposes. First, it is a standard diet to which many experimental diets are compared; second, because we have had much experience with the casein diet, it would provide a check on whether something was amiss with the batch of rats or with some other aspect of the study. Things sometimes go wrong in mysterious ways, and it is important to have some sort of check.

At the start of the experiment, the 30 animals used varied in weight from 50 to 63 grams. They were arranged in ascending order of weight, and from the three lightest ones, one was assigned to each of the three diets in the assay. Such a trio of approximately equal weight animals is called a *block*. The assignment of rats to diets within this block of three was by chance or at random; that is, random numbers determined which rat went on each diet. The second block of the next three lightest rats was assigned one to each diet in the same way. This process of randomized block assignment continued until all 30 rats set aside for this study were distributed among the three diets, 10 per diet.

Using blocks of rats of comparable body weight ensures that each diet has its proper share of light and heavy rats. Randomization helps balance other factors which may influence the outcome of the study. For example, the shipment of 30 rats probably included many who were littermates: they were all newly weaned male rats and so were all born at approximately the same time. The probability is high that the randomization spread the rats from one litter among the three diets. Of course, to be sure that a litter is equally divided among the three diets it would be necessary to identify the rats by litter, reduce a litter size to a multiple of three by random withdrawal of animals, and randomly assign the remaining rats equally to the three diets. This would be done if some inheritable trait might importantly affect the results of a study.

The rats were assigned to cages by random numbers because prior studies had shown that rats at the top of a rack of cages gain more weight than those at the bottom. This randomization guarded against the possibility that most of the rats on one diet were put in top cages while rats on another diet went into bottom cages.

During the H feeding study each rat was permitted to eat as much as he wished. His food consumption was measured, and his weight was checked weekly. The final weight gain was simply the difference between his final, 28-day body weight and his initial weight. Intermediate weights are used to check on any untoward event such as respiratory infection, since, in the rat, body weight is a sensitive indicator of general well-being. The total food consumed times the proportion that is protein—a value obtained by chemical analysis of samples from each diet—gives a rat's protein intake. For the H study, the results for these two measures are shown in Figure 1. Each point on the graph represents one rat. All three diets were made up to have about the same proportion of protein, 9%. The chemical analysis showed them to be close to that: 9.875, 9.875, and 9.50%. Therefore, the higher intake of protein for rats fed the H diets means they ate more food which may indicate that these diets were more palatable to the rats than the casein diet. However, experienced nutritionists claim that rats generally will eat heartily any balanced high protein food irrespective of texture and flavor. All that is known from the data is that the intake of H was substantially above the intake of casein. The other most obvious fact about the points in Figure 1 is that they tend to fall close to a slanted straight line drawn through the middle of them. However, closer inspection of Figure 1 with ruler in hand indicates that a freehand straight line through liquid H points lies slightly above a line through solid H points and has a steeper slope. A line through the casein points alone would lie below either of the H lines, if it were extended to the higher intakes. Thus some doubt was cast on the initial conclusion that all points fell on one line and thus on the hypothesis that any differences between the dietary weight gains could

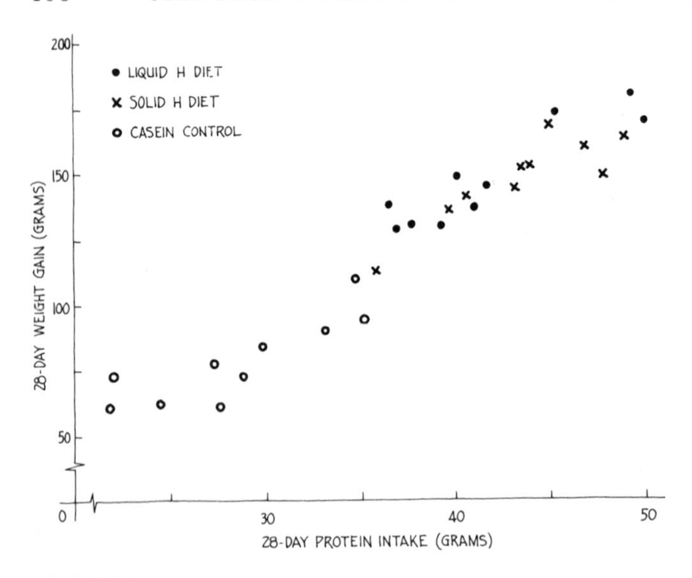

FIGURE 1
Relationship of 28-day protein intake and weight gain in young male rats

be attributed entirely to differences in protein intake; that is, we began to think that the difference in weight gain was not due simply to the difference in protein intake.

Using a computer program, we fitted a straight line to the points for each diet by the method of least squares; that is, the line was selected which minimized the sum of the squared vertical distances of the points from the line. These fitted lines, called *estimated regression lines,* are described by equations of the form $Y = \bar{Y} + b(X - \bar{X})$, where Y is the predicted weight gain and X the protein intake. $\bar{X}$ *and* $\bar{Y}$ are the averages of the X's and Y's separately for each set of data, and each b represents the slope of that straight line, that is, the predicted increment in weight gain per unit of protein intake.

The three regression equations were:

Liquid H: $Y = 151 + 3.72(X - 41.7)$
Solid H: $Y = 150 + 3.66(X - 43.3)$
Casein: $Y = 79 + 2.91(X - 28.4)$.

Inspection of the three lines—they are graphed in Figure 2—shows that the casein line is decidedly lower and slopes less than the two H lines. The two H lines were then compared, and we found that they did not differ more than one would expect by chance. Thus our suspicion was confirmed that the greater gains of the H fed rats relative to the gains for rats on

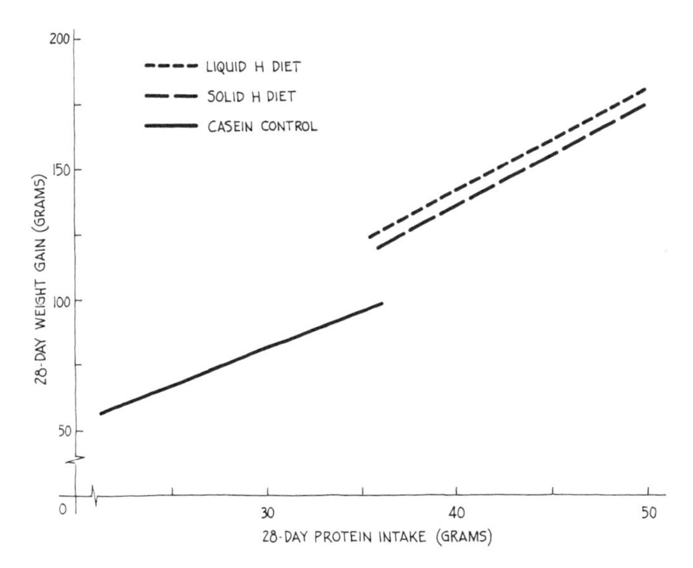

FIGURE 2

Estimated regression of 28-day weight gain on protein intake for young male rats

casein were not due simply to greater protein intake, because the slope of the H lines is higher.

To summarize the results of the protein evaluation: differences in protein intake between the casein fed rats and H fed rats did not account for all the differences in weight gain. For a given increase in protein intake the H diets resulted in a greater increase in weight gain than did the casein diet. There was little difference between solid and liquid H in terms of expected weight gain. If the addition of solidifying agents is detrimental it was not evident in this assay.

PALATABILITY EVALUATION

While the study to determine and compare the nutritive values of the two H variants was being run, a preliminary consumer test to determine the palatability of the two variants also was run. The aim was to compare the two H products and to see if they were as acceptable overall as a competitive product already on the market, designated C.

Previous experience had shown that having 50 people taste and evaluate a product under controlled conditions is adequate to reveal a major problem in acceptability, if one exists. Controlled conditions were obtained by bringing individuals into a central testing location and having trained personnel prepare, serve, and interview the 150 individuals required—50 to taste C and

FIGURE 3
Pictorial taste-test ballot (the scores +3 to —3 were assigned the figures from left to right)

50 to taste each of the two variants of H. Because men and women might respond differently, the test specified that each product would be tasted by 25 males and 25 females. The tasters, who were paid for their help, were recruited from local churches or club groups. As they were recruited, individuals were assigned randomly to taste each of the three products until 25 of each sex had tasted and evaluated each of the three products.

After tasting the product, each person was asked to mark a ballot rating overall acceptability. Figure 3 shows the pictorial scale used to measure acceptability. Individuals seem to find it easier to express their feelings about a product using a pictorial scale rather than a scale using words such as "excellent," "very good," "good," "average," "poor," "very poor," and "terrible."

Each of the pictures is assigned a number (or score) in the sequence +3, +2, +1, 0, —1, —2, —3, with +3 being most acceptable.

For each of the three products the total number of votes for each of the pictures was tallied separately for males and females. The results, called frequency distributions, are shown in Table 1 with the pictures being replaced by the assigned number or score, and M being for males and F for females.

From the frequency distributions it is apparent that all those tasting the same product did not agree on its acceptability. It is difficult to look at the six distributions and decide whether any one group of 25 is scoring what they tasted as more or less acceptable than each of the other five groups scored what they tasted. To simplify the comparison among the six groups of 25 people an average score is obtained by multiplying each score by its frequency, summing the results, and dividing by 25.

We note that the differences in average scores are not large when male tasters are compared to female tasters. We note too that the averages for

TABLE 1. Frequency Distribution of Scores

	PRODUCT TASTED					
	C		**Liquid H**		**Solid H**	
SCORE	M	F	M	F	M	F
+3	1	0	4	3	2	3
+2	2	3	6	7	6	5
+1	7	8	9	7	10	9
0	8	9	5	6	6	7
−1	5	3	0	2	1	0
−2	2	1	1	0	0	1
−3	0	1	0	0	0	0
Total tasters	25	25	25	25	25	25
Average score	0.20	0.24	1.24	1.12	1.08	1.04
Variance	1.50	1.43	1.44	1.36	0.993	1.2

liquid H and solid H don't differ much, but all four of the H averages are well above the two C averages. The question to be answered is whether these differences in average scores are larger than can be expected by chance, considering the way people vary when they rate the same product. What is needed is a yardstick to permit us to say what difference between any two averages, each based on 25 tasters, is larger than can be expected by chance alone.

To arrive at such a yardstick we must first measure how variable people within each of the six groups are. This measure, called *variance,* is obtained by taking the difference of each person's score from the average, squaring the differences, summing the squares, and dividing by one less than the number of people. For each of the six groups the variance was computed; it is shown at the bottom of the table of frequency distributions. The variances for all six groups in this study were then averaged to give 1.32, a good measure of variability among males or females tasting the same product

If individuals vary, so will the averages based on individuals. How much the averages vary will depend on the number of tasters: the larger the number the more representative and less variable the average will be. Because of this variability we can never be absolutely sure that two averages are different, but we must take some risk in drawing conclusions. So the best we can do is state the risk in setting up our yardstick. We chose to take a "1-in-20" chance of being wrong when we calculated the amount by which two averages had to differ in order for us to conclude they were different. Using the variance of the average of 25 scores and taking into account the risk gives 0.64 as our yardstick for comparing any two of the average scores.

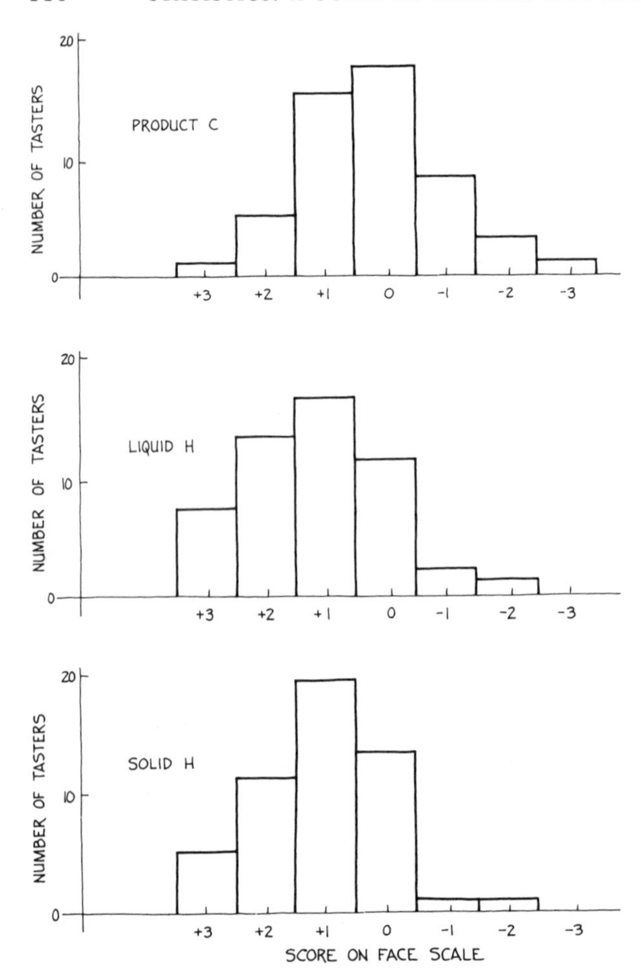

FIGURE 4
Combined male and female distributions of face-scale scores by product tasted, 50 tasters per product

This tells us that there is no evidence that males and females who rate the same product rate them differently because we have for the differences in averages the following:

C:	$0.24 - 0.20 = 0.04$
Liquid H:	$1.24 - 1.12 = 0.12$
Solid H:	$1.08 - 1.04 = 0.04$.

We can now combine the distributions for males and females. The resulting distributions are shown in Figure 4. We note that the distribution for product C is shifted to the right, the less acceptable scores, as compared to the H results. There is considerable overlapping of the distributions representing the variation, so even on a pictorial basis we have problems interpreting the relative acceptability of the test products. We must resort to the average scores, which for the 50 individuals who tasted the same product are the following:

C = 0.22 Liquid H = 1.18 Solid H = 1.06.

Our yardstick for judging differences decreases, because these new average scores are based on 50 people. Our new yardstick, taking a 1-in-20 chance of being wrong, is 0.45.

It is apparent the difference between the acceptability of the two H products is not large enough to say one is more acceptable than the other. There is very little risk, however, in concluding that both H products are more acceptable than C, the competitive product.

Thus, through testing and with the application of statistical methodology these new products were shown to be palatable and to live up to the concept of a high-protein food. Two of the early criteria in the long process of introducing a new food product have been met.

PROBLEMS

1. For the protein evaluation, why were rats chosen as the experimental animals?

2. (a) Explain the idea of blocking.

(b) Suppose the rats had not been assigned via blocking but instead were assigned in the following way: For each of the thirty rats a six-sided die is rolled. A 1 or a 2 means the casein diet for that rat, 3 or 4 liquid H, and 5 or 6 the solid H diet. Under this scheme how many rats would be assigned to each diet? Would it be possible that all rats are assigned to the casein diet?

3. In the protein evaluation experiment, what was the least protein intake for any rat on the casein diet? The most? Answer the same questions for the liquid H diet and the solid H diet.

4. What was the median weight gain of the 30 rats in the protein evaluation experiment?

5. Using the appropriate equation derived from the protein evaluation experiment, how much weight would you predict a rat would gain in 28 days if he consumed 45 grams of protein on the casein diet? State the assumptions that led you to this prediction.

6. Suppose you were setting up the protein evaluation experiment. Explain how you would use the random number table on page 88 to
   (a) assign rats in each block to the three different diets;
   (b) assign rats to the top or bottom cages.

7. (a) The yardstick for comparing average scores between males and females tasting a food in the palatibility evaluation was .64; for comparing average scores between two different food products, it was .45. Why are these yardsticks different?
   (b) Do you think you could ever achieve a yardstick smaller than .1? If your answer is yes, how might you do this?

8. Calculate the variance of the following scores for a product B.

SCORE	FREQUENCY OF SCORE
+3	2
+2	7
+1	7
0	7
−1	2
−2	0
−3	0

TOTAL TASTERS 25
AVERAGE SCORE 1

INFORMATION FOR THE NATION FROM A SAMPLE SURVEY

Conrad Taeuber *Associate Director, Bureau of the Census*

THE DOORBELL at the Robert Brown house rings as the Brown family is finishing lunch. Nick Brown, the oldest son of the family, goes to the door. "Mom," he calls back, "it's Mrs. Smith, the Census Bureau lady" This is not the Census lady's first visit to the Brown's, so no further introduction is needed. "Well, tell her to come on in," says Mrs. Brown, and Mrs. Smith is invited to join the Browns for coffee.

Courteously, but without wasting time, Mrs. Smith verifies that all members of the Brown family are still living at home and that no one is there to visit or stay. Assured that there are no changes, she begins her questions:

> Mr. Brown, what were you doing most of last week, working or something else? How many hours did you work last week at all jobs? For whom did you work? What kind of business or industry is this? What kind of work were you doing? Mrs. Brown, what were you doing most of last week, keeping house or something

> else? Did you do any work at all, not counting work around the house? Did you have a job or business from which you were temporarily absent or on layoff? Have you been looking for work?

Mrs. Smith turns to Nick, who says he's been looking for work:

> What have you been doing to find work? Why did you start looking for work? How many weeks have you been looking for work? Have you been looking for full-time or part-time work?

And similar questions are repeated for each member of the Brown household who is 14 years old or older.

Every month, about 50,000 families in the nation are interviewed in this way. Yours may have been one of them, although that is unlikely. Why is this program carried out? To understand this, we turn from the specific interview in the Brown household to the broad national scene.

A newspaper headline in the fall of 1969 announced "U.S. Jobless Rate Advances to 4 Percent, Highest Since '67." The same newspaper went on to observe, "Administration aides greet rise as sign of restraint on boom spurring inflation. Former Vice President [Humphrey] is reported as being critical, but a Treasury official said no change in tax policy is yet under consideration." The reporter had learned that the development was greeted by Administration officials as a welcome sign that their policies, aimed at ending inflation by slowing business expansion, were effective.

The rise in unemployment that month was spread throughout the work force. Unemployment rates increased for almost every category of workers—adult men, adult women, teenagers, Negroes, and Caucasians. The ones most seriously affected, however, were men in the 20 to 24 age group, blue collar workers, construction, trade and manufacturing employees, and agricultural workers.

These facts and many more gave the nation a major indication of the state of our economy in September 1969. Similar figures are available every month and they provide the Congress and the Administration with essential information to help in charting the course of the economy. The figures are widely published. Both the Administration and its critics find in these figures information about the levels of employment and unemployment and the persons who are most directly affected by changes in employment. At a time when there is serious concern with the possibility of inflation and with the consequences of efforts to slow down the rate of economic growth, the statistics on unemployment are carefully watched. Responsible officials want to be sure that the policies are working as desired and that the Government has an early warning if the effects are so great that action is needed to try to avoid dangerous levels of unemployment or recession.

Such statistics, and many others, are essential tools in the nation's efforts to guide its economy. Information is needed on how many people are working, how many are unemployed, whether the workers are on overtime or on short time, whether the unemployed are primarily married men, single young men or women, and whether unemployment is hitting harder at young or older ages and at black or white. There is a need also for information on how many workers are on temporary layoffs, how many are young persons looking for their first jobs, and how many are women looking for jobs when their husbands have been thrown out of work.

Unemployment may mean many different things, depending on who is unemployed and for how long. The unemployed worker may be simply between jobs, with a short break before he takes on the new one. He may have been laid off because a plant has been shut down or relocated, or simply because a plant has temporarily reduced its work force. He may be a young person testing the labor market, looking for a job for a short time, or trying to decide whether to get a job or to stay in school. She may be a housewife who would like some additional work to help out with the family expenses, to buy some special item, or to finance the schooling of the children. She may be a housewife needing a job in order to supplement a husband's inadequate income or to make up for the fact that he has no income. The unemployed worker may be a man who finds it difficult or impossible to leave his home area where there is no job in order to go to some other location where one may be available. He may be an older worker whose skills have become obsolete and who faces continuing unemployment unless he can be retrained. Among three million unemployed, there are people in many different situations. Because of its concern with the welfare of all Americans and its commitment to a policy of full employment, the U.S. nation has developed a wide variety of measures to help meet the problems of unemployed workers. It needs a continuing source of reliable information on the actual situation at a given time and on the changes from one month and one year to the next.

We now have much more up-to-date information than we did in the early thirties when the nation was deeply concerned with the problems of unemployment. Then everyone knew that there was an unbearably large number of unemployed, but hard information was lacking on how many there were and what types of persons they were. Different authorities made widely differing estimates. No one had a real count; people had to rely on guesses. Though the Administrator of the Federal Emergency Relief Administration could state emphatically that hunger is not debatable, even he did not have a good measure of the size or nature of the problems that he had set out to correct. It was the early efforts of the Works Progress Administration to get some reliable measure of changes in the levels of unemployment that led to the establishment of the survey which now provides the official monthly figures on employment and unemployment.

These figures, of course, do not settle the arguments about how much unemployment is normal, tolerable, or dangerous. Nor do they entirely satisfy the people who feel that a person on part-time work, but who would like full-time employment, should be counted as unemployed. Some think that 14- and 15-year-olds who want work should be counted as unemployed. There can be differences of opinion about the proper classification of persons on temporary layoffs, those who are not working because of some labor dispute, seasonal workers in the off-season, the "discouraged workers" who have given up looking for a job because they are sure none is available, and the persons who aren't sure but say they would like a job if an attractive one comes along. It is sometimes argued that the primary concern should be about married men with dependents and that social economic policy need not be seriously concerned about others who may be unemployed.

Statistics cannot settle such policy questions, but they can supply information that helps in identifying essential issues and in narrowing the debate to the policy issues. Persons advocating different policies can carry on useful discussions or debates only if they are agreed on certain basic propositions. If they can agree on the statistics, they can then match their different interpretations against each other. If, however, they do not agree on what the basic facts are, much of the argument may be fruitless. Even though they may disagree on precisely who should be included and who should not, they can narrow the policy argument if they can agree on how many and what kinds of people are included in the groups about which they disagree.

How do we know about the number of unemployed and whether they are young or old, men or women, black or white, blue collar or white collar workers? How do we know how many of them have just become unemployed and how many have been looking for work for three months or more? How do we know that unemployment in one month is more or less than it was in the preceding month or in the same month a year ago? There are important seasonal changes in some kinds of work—the harvesting of crops, canning, retail selling with its large need for temporary help before Christmas and Easter, and many others. We need to take all of these seasonal changes into account in assessing whether a change from one month to the next is really significant or whether it is simply a reflection of the normal seasonal development.

THE CURRENT POPULATION SURVEY

The basic source of such information, both in this country and in some others, is the Current Population Survey (CPS). In the U.S., every month some 1100 persons leave their housework or other duties and call at some 60,000 addresses to ask the occupants a number of carefully worded questions. The answers (usually there are some 50,000 of them) are promptly assembled, and from them, the nation has its monthly report on employment and unemployment.

Every month, in the week that includes the 19th, these interviewers begin their rounds. Each one has about 60 addresses on his or her list. The questionnaires are so designed that many of the answers can be recorded simply by filling in little circles. Some of the answers, however, must be written out, for example, the kind of work that each worker was doing. The interviews begin on Monday and always relate to activity during the previous week, that is, the week that includes the 12th of the month. Completed questionnaires go to a regional office of the Census Bureau, where they are given a quick review, and then promptly shipped to the Census Bureau's processing office in Jeffersonville, Indiana. There trained persons translate the answers from ordinary language into a language that the computer can read. If the interviewing begins on the 15th of the month, all of it is completed by the 22nd, and the computer receives all of the forms by the 29th. Overnight it does the work of combining the data into usable tables. Approximately two weeks after that month's enumeration began, the first statistics are available. They need to be reviewed to be sure all of the processing, including that done by the computer, has been done in accordance with the instructions. Within two days more, the results are ready for distribution to newspapers, radio, and television, which pass on the information to the public. Analysts in the Government and outside then have a new set of figures to analyze in order to arrive at proper policy decisions in the light of the changes revealed that month. This process is repeated month after month, as it has been in essentially similar form for more than 25 years.

Anyone using such statistics is likely to ask how reasonable it is to draw conclusions about all of the 60 million households in the country from a sample of a little more than 50 thousand (about 0.1%). To answer this natural question, we must look at the method by which the sample of households is drawn. If that method is highly restrictive (for example, if only families in major industrial cities were selected or only farm families), then it would obviously be inappropriate to reach conclusions about all the country from the sample. Similarly, if families were chosen by asking each interviewer to interview 50 families who are his friends or neighbors, we might properly doubt inferences from the sample. We want to be sure that the sample has appropriate numbers of rural and urban families, of high- and low-income families, and so on. All parts of the country should be represented in proportion to their share of the population.

The major assurance that the sample will reflect conditions for the entire country comes from the fact that the households are probabilistically selected from within homogeneous strata. Not only does this help to keep the process free from personal bias on the part of the persons choosing the sample, but it also permits accurate knowledge of the precision of the estimates that come from the sampling. This also means that it is possible to determine the sample size that will assure the degree of precision in the data needed to meet the requirements of public policy. The details of how the sample is selected are described below.

There are more than 60 million households in the U.S., and the problem is how to select an appropriate sample each month of about one out of every 1240 for interview. If there were an up-to-date list of all households in the U.S., one out of every 1240 of them might be selected in an appropriate manner, but there is no such list.

Because the information is to be collected by interviewers visiting the households, it is desirable to select the sample in clusters of neighboring households. This helps the interviewer get to all of them within the few days each month in which the interviewing must be done. In this survey, clusters of about six households are used.

Another requirement is that the sample should be selected in such a way that it is possible to estimate from the sample itself how much the results differ from those which would have been secured if every household in the country had been included. It is well known that the results from a sample are rarely precisely the same as those from a complete enumeration, but from a properly designed sample, we can measure the chance that these deviations are small enough so that the major findings from the sample can be used with confidence. The same type of consideration applies to the comparison of changes from month to month. For them, too, it is necessary to know whether a difference is real or is simply the kind of chance difference that could be the result of "sampling error."

The problem for the statisticians was to develop a way of drawing the sample from the entire country in such a way that an economical and reliable survey could be conducted each month. A measure of the degree of confidence to put in the figures also should be obtained from the sample.

It was decided to give each interviewer a fixed set of addresses at which to call. If no one is living at one of the addresses or if the persons found there do not actually live there, that address will not contribute data to the survey for that month. If a household moves away from an address between two of the monthly visits, it is dropped from the survey and the household which has moved to the sample address is included. The interviewer is responsible for completing an interview at each of the assigned addresses, or explaining why no interview was required or possible. If no one is found at home after several calls or if the residents at the assigned address refused to provide information, it is so noted. No substitutions for sample households are allowed.

The Current Population Survey, which is the source of this information, is carried out in 449 sample areas, which include 863 counties and independent cities. They are located in every state and the District of Columbia. In all, about 60,000 residential addresses are designated for the sample each month; about 52,000 of them, containing about 105,000 persons 14 years of age and over, are actually interviewed. The survey is limited to the civilian population and excludes members of the Armed Forces, as well as persons living in institutions, such as prisons, long-stay hospitals, and so on. Most of the 8000 addresses for which interviews are not secured are for housing units that are vacant, units that have been converted to nonresidential use, units whose usual occupants are temporarily away, or those that are temporarily occupied by persons who actually live somewhere else (for example, persons occupying a home on their vacation). Answering the questions is voluntary, but fewer than 2% of the persons interviewed refuse to answer.

The first step in selecting the households is to select a sample of the counties or equivalent governmental units. These are then subdivided into subareas, and a selection is made among these subareas. Within each of the selected subareas, a sample of addresses is selected for interviewing.[1] In making the final selection of the specific addresses to be used, two different procedures are utilized. In urban areas it is generally possible to work with addresses that give specific house numbers and streets and even apartment numbers. When such lists are available, a cluster of about 18 consecutive addresses is selected from the census enumeration districts (ED). Every third address within the cluster is taken for the current sample; the remaining 12 addresses are saved for use in future samples. Arrangements are made to take into account new construction since the last census, chiefly by checking building permits.

In rural areas and other areas where such addresses are not available, *area sampling* is used. The sample EDs are subdivided into small land areas with well-defined boundaries. Insofar as can be determined from available information, each such area segment has about six housing units. If it is not possible to define area segments of that size, larger segments may be defined. These six addresses are drawn by a systematic sampling of all housing units.

It is desirable to have a household in the sample for consecutive months, and for the same months in successive years in order to secure measures of month-to-month and year-to-year change. To avoid overburdening the households who cooperate in the interview, it has been arranged that interviews are conducted at a sample address for four successive months, then that address is omitted for eight months, and after that, it is interviewed again for four consecutive months. This rotation occurs in such a way that each month one-eighth of the sample addresses are entirely new, one-eighth consists of addresses that are starting on the second round of four interviews, and three-

[1] See the appendix to this essay for a more detailed description of the sampling procedure.

fourths were interviewed in the preceding month. Thus one-half also were interviewed in the same month a year before.

PRECISION AND ERROR CONTROL

Modern sampling theory makes it possible to measure the size of the fluctuations arising from the sampling process, for the probability of including any unit in the sample is known. Of course, as in any survey, there are also other, nonsampling sources of error that must be investigated and controlled. Interviewers must be carefully trained both in the techniques of interviewing and in the content of the questionnaire they are using. Every effort must be made to assure that respondents understand the questions and provide correct answers. Controlling such a series of interviews requires attention to all possible sources of error, and taking appropriate steps to control them. In the case of the Current Population Survey, there is continuing training of the interviewers, careful review of their work each month, periodic observation, and a program of reinterviews by supervisory personnel.

On the average of twice a year, a subsample of the addresses assigned to each interviewer is visited a second time by a supervisor and the same questions are asked again to make sure that the correct information has been obtained. The interviewers do not know when their work will be checked or which addresses will be selected for the reinterview. The supervisors do not know at which addresses they are to reinterview until after the initial interviews have been completed. If the information secured at the two interviews differs, an effort is made to find out which of the two answers is correct and why the differences occurred. The reinterview program serves as a basis for further training of the interviewers and gives a measure of the quality of the survey in general.

All steps in the office processing of the interviews are similarly kept under constant control. The preparation of the estimates in which the results of the sample are projected to the entire population also requires application of modern statistical principles. The results are published with a measure of the sampling variability of each of the major figures.

QUESTIONS ASKED

Selecting the sample properly and training and supervising the interviewers carefully to make sure that they interview at the designated addresses and report the replies accurately would not be adequate if the questionnaire itself were not well designed or if the questions to be asked were left to the interpretation of each interviewer or each respondent. A question such as "Were you unemployed?" would give results of little value, for there would be wide differences in the interpretations placed on such a question. Instead, over the years a battery of questions has been developed to secure information on what a person actually did during the survey week and to classify him or her on that basis. If he reported that he was working, he is asked how many hours he worked during the week, and because people are likely to report some standard number such as 40, he is also asked whether he worked any overtime or lost any time or took off any time. He is asked also to state what his job was and for whom he worked, as well as the kind of business or industry in which he worked.

If he did not work during the survey week (i.e., the week preceding the interview) he is asked whether he had a job or business from which he was temporarily absent or on layoff. If he was absent, he is asked why and whether he was paid for the time off. If he was not working, but looking for work, he is asked to indicate what he has done during the preceding four weeks to find work, such as checking with an employment agency, with employers, or with friends and relatives, placing or answering advertisements, and so on. He is also asked why he started looking for work, whether because he lost his job, quit his job, left school, or wanted temporary work. There are questions to ascertain how long he has been looking for work, whether he was looking for full-time or part-time work, and whether there was any reason, such as illness or school attendance, why he could not have taken a job last week. There are questions on when he last worked for pay, and the kind of work he did on his last job.

If he is not looking for work, he is asked when he last worked for pay, why he left his last job, and whether he intends to look for work of any kind in the next 12 months. He is also asked why he is not looking for work, whether because he believes that no work is available in his line of work or in the area in which he lives, that he lacks the necessary training or skills, that employers think he is too old or too young, or that there is some personal handicap that stands in his way. Other possible reasons include family responsibilities, the inability to arrange for child care, ill health, a physical disability, or the fact that he is going to school. The answers to these questions provide a basis for meaningful classification of persons as employed or unemployed. Information on the respondent's sex, age, color or race, marital status, relationship to the head of the household, and years of schooling is also available to assist in providing some meaningful classification of persons as employed, unemployed, or not in the labor force.

The published results provide a monthly measure of national employment and unemployment. From time to time a few questions are added to the questionnaire. These special questions, in addition to those that are asked every month, provide the basis for much of the information about important social trends in the years between the major censuses.

Because the sample continually reflects the growth or decline of population in the sample areas, it provides a basis for estimating the major geographic shifts of the population. Between 1960 and 1968 the total population of the central cities of our metropolitan areas remained almost the same, but there was a major change in the makeup of the population of these cities, including a net loss of about 2.1 million Caucasians and a gain of about 2.6 million Negroes. The continuing movement of people to the suburbs is clearly reflected in the statistics from this survey, with about four-fifths of the national growth occurring in the suburban areas. The survey has also reflected a slowing down of interregional migration during the sixties, compared to the higher rates of migration in the fifties.

The annual report on the number of families and persons in poverty comes from this same survey, for once a year each person in the survey is asked to report the amount of income he received during the preceding year. The number of persons living in poverty declined from nearly 40 million in 1959 to about 25 million in 1968. The percentage of the population living in poverty declined from 22 in 1959 to 13 in 1968. There were differences in regard to income and poverty between Caucasians and Negroes, as shown by these figures. In 1959, more than half the Negroes were classified as living in poverty; in 1968, the comparable figure was 35%. Average family income has been increasing. In 1968, it amounted to $8600, which represented a real gain of about 3.5% over the previous year, even after allowing for the increased level of prices.

An important measure of educational attainment is the proportion of young adults who have completed high school. Between 1960 and 1969 the proportion of white men 25 to 29 years old who had completed at least four years of high school rose from 63 to 78%. For black men in the same age bracket, the percentage went up from 36 to 60% in the same nine years.

Americans are a mobile people, and the survey provides a measure of that mobility. One person in five changes his residence in the course of a year; most of these moves are within the same county. There has been very little change in this rate of mobility over the last 20 years for which information is available.

The survey also indicates that in homes with children that include only one parent (usually the mother) the family income is likely to be relatively low. In 1969, 89% of white children and 59% of black children were living with both parents. In recent years, there has been little change in this percentage among whites, but some decrease among blacks.

Such facts about life in the U.S. are available annually because the families included in the survey are willing to answer the questions put to them by the interviewers. A census covering the entire population is taken only once in 10 years. Between censuses, we now have statistics that reveal changes for the nation as a whole and for the major regions. A survey of this size, however, cannot provide figures of the same reliability for states, individual cities, or for metropolitan or smaller areas.

From time to time, the Bureau of the Census is asked to add some other questions to the questionnaire. As a result, it has been able to supply statistics on the number of persons who smoke, and the proportions of young men and women and older men and women who do so, as well as the numbers who quit smoking. The most recent survey showed that about 2.7 million persons had quit smoking between 1966 and 1968. Some persons began smoking during that time, but the total number reporting that they smoke dropped by about 1 million. Information has also been supplied on the number and proportion of persons who have had immunization for polio, smallpox, diptheria, and other communicable diseases.

In view of the public interest in the percentage of persons who vote, questions have also been asked about whether the individual had voted and, if not, whether he had been registered to vote. The survey found shortly after the 1968 election, that 68% of all persons of voting age reported that they had voted. Men had higher voting participation rates than women. Northern blacks had a higher voting rate than Southern whites. Persons over 65 and those under 35 had lower voting participation rates than persons between 35 and 64; persons with higher educational levels and those with higher incomes tended to have higher voting participation rates. Unskilled workers had lower voting participation rates than persons in the occupations that required higher levels of training.

THE ROLE OF THE SURVEY

This survey, which involves the willing cooperation of more than 50,000 households each month, has proven to be an important source of information about the U.S. This information is needed by the Government in planning its economic and other policies. It is also needed by many other agencies and by private organizations concerned with employment and unemployment, educational levels, poverty, incomes, health, and living conditions generally. Through the application of modern sampling theories and through a system of carefully developed training and supervision, it has become possible to provide this and much other information on a timely basis and at a fraction of the cost of conducting a complete census. Congress and other policy makers and administrators look to this source for up-to-date information, knowing that the results are reliable.

Such a survey is not a substitute for a census, which provides information for each state, county, and city, and for smaller areas, but it does provide essential information between censuses and is an indispensable source of data needed for a continuing appraisal of important developments in the nation.

APPENDIX: PROCEDURE FOR SELECTING SAMPLES FOR THE CURRENT POPULATION SURVEY

The first step in selecting the addresses to be visited was to determine the counties in which the interviewing was to be done. This was accomplished by combining all of the counties in the U.S. into Primary Sampling Units (PSUs). Each of the Standard Metropolitan Statistical Areas (SMSA) was taken as a PSU. An SMSA is a city of 50,000 or over, plus the county in which it is located and adjoining counties that are closely tied to it economically and socially as determined by certain specified criteria. Outside the SMSAs, counties were grouped into PSUs (small groups of counties that are sufficiently compact so that a sample of households within a unit could be visited without undue travel cost). Whenever possible, a PSU was constructed to include both urban and rural residents of both high and low economic levels and, to the extent feasible, a variety of occupations and industries. When the current sample was selected in 1962, there were 1913 PSUs, including the 212 SMSAs defined in the 1960 census.

The next step was to combine these PSUs into 357 strata. Each of the SMSAs with 250,000 or more residents was treated as a single stratum. The other strata, in general, consisted of sets of PSUs as much alike as possible in various characteristics. These included geographic region, density of population, rate of growth between 1950 and 1960, the proportion of the population in 1960 that was not white, the principal industry, type of agriculture, and so on. Except for the strata in which an area represented itself (such as the larger SMSAs), the strata were so arranged that their populations in 1960 were approximately equal.

If a PSU was a stratum by itself, it automatically fell into the sample, and thus there were 112 strata that consisted of only one PSU. Within the other 245 strata, PSUs were selected for this sample at random in such a way that the probability of each being drawn was directly proportional to its 1960 population. Thus, within a stratum, the chance that a PSU with 100,000 persons would be drawn was twice as great as that of a PSU with only 50,000 persons.

The next step was to select the sample households within the designated PSU. The sampling rate within each PSU was determined in such a way that the overall chance of an address being included is equal to one in 1240. In this way, it is possible to reflect changes, such as new construction or demolition of housing units or changes in the characteristics of the population of the area that have occurred since the last census was taken. Within each designated PSU, the first step was to select a sample of the census enumeration districts (ED) that were used in 1960. These small administrative units contained approximately 250 households in the 1960 census. The EDs were arranged in geographical order to make sure that the sample EDs will be spread over the entire PSU. The probability of selection of any one ED is proportionate to its 1960 population.

PROBLEMS

1. Explain why a large battery of questions are asked on unemployment rather than just the question, "Were you unemployed?"

2. Briefly explain how households are selected for inclusion in the CPS.

3. List at least three advantages which the Bureau of the Census gains by sampling 60,000 households rather than interviewing every household in the Current Population Survey.

4. Are the figures reported monthly for unemployment among black males aged 20–24 likely to be just as accurate, more accurate, or less accurate than the figures reported monthly for unemployment among all workers? Why?

5. Why doesn't the Census just sample employers and find out how many workers are employed, instead of carrying out the CPS to determine an unemployment index?

6. The Census uses rotation sampling instead of taking a completely fresh sample every month. Why? Do you think that the Gallup Poll should use rotation sampling? What about the Nielsen television ratings survey?

ACKNOWLEDGMENTS

We wish to thank the following for permission to use previously published and copyrighted material:

Interstate Commerce Commission, Bureau of Economics, for permission to reprint the table on p. 88 from its *Table of 105,000 Random Decimal Digits,* 1949.

Simmons-Boardman Publishing Co. for permission to reprint the table on p. 90, from "Can Scientific Sampling Techniques Be Used in Railroad Accounting?" *Railway Age,* June 9, 1952, pp. 61–64.

INDEX

47-301